Building Internet of Things with the Arduino

V1.1

CHARALAMPOS DOUKAS

ISBN: 1470023431
ISBN-13: 978-1470023430

DEDICATION

To my family for making me what I am, to my professor and mentor Ilias Maglogiannis for sharing with me the knowledge and the challenges and to my Zafi for beautifying my life.

CONTENTS

Acknowledgments   i

Introduction   1

1   The Internet of Things   8

2   The Basics of Sensors and Actuators   25

3   Introduction to Cloud Computing   42

4   The Arduino Microcontroller Platform   63

5   Reading from Sensors   103

6   Talking to your Android phone with Arduino   122

7   Connecting Arduino to the Internet   156

8   Introducing Cosm as a Cloud Service   213

9   Introducing the Nimbits Public Server Cloud   269

10   Reprogram your Arduino remotely from the Cloud   291

11   What you can connect to the Cloud: Project Ideas   311

CHARALAMPOS DOUKAS

ACKNOWLEDGMENTS

Firstly, I would like to thank CreateSpace for giving me the opportunity to publish this book. I would also like to thank Sparkfun and Seeedstudio for letting me use many of their product images in the book chapters. Special thanks to people from Cosm (former Pachube) and Benjamin Sautner from Nimbits for developing and providing such great web tools for the IoT. Many thanks to the Jelastic team for letting me host projects on their so-easy-to-use Cloud platform. The Arduino team deserves a great acknowledgement for providing the open hardware community with such a great tool! Last but not least, I would like to thank the open hardware and software community and the anonymous contributors of the various Arduino libraries that have made Arduino programming and communicating with the Internet so accessible!

"The future is already here — it's just not very evenly distributed."

William Gibson

INTRODUCTION

Have a look around you and try to identify devices that can sense something about their environment. No-matter if you are at work, at your home, or next to a bookshelf, I am sure you will identify many: Temperature sensors, Gas sensors, Thermostats, infrared sensors (used in automatic doors), etc. How many of these devices can communicate with the Internet and interact with Web-based applications? Probably none or just a few of them, but this is soon going to change. The evolution of communication technologies is bringing Internet connection to devices at lower cost, less power consumption and smaller sizes, making devices able to be parts of the so called 'Internet of Things': a global network of smart devices that can sense and interact with their environment using the Internet for their communication and interaction with users and other systems.

The Arduino is a programmable device that can sense and interact with its environment. It is a great open source microcontroller platform that allows hobbyists and electronic enthusiasts to build quickly, easily and with low cost small automation and monitoring projects. It has been so widely adopted that dozens of vendors build several variations of it and various extensions that provide features like wired and wireless connectivity to the Internet. In addition, there is great software support by the open source community: coding an embedded device has never been so easy.

I consider Arduino the best way to be introduced into the Internet of Things (IoT) concept. 'Building Internet of Things with the Arduino' aims to give you exactly all the knowledge you need to start with your own IoT projects.

What You Will Learn

This book will provide you with all the information you need to design and create your own IoT applications using the Arduino platform. More specifically, you will learn:

- About the Internet of Things and Cloud Computing concepts
- About open platforms that allow you to store your sensor data on the Cloud (like Cosm and Nimbits)
- The basic usage of Arduino environment for creating your own embedded projects at low cost

- How to connect your Arduino with your Android phone and send data over the Internet
- How to connect your Arduino directly to the Internet and talk to the Cloud
- How to reprogram your Arduino microcontroller remotely through the Cloud

How The Book Is structured

The Book is divided into three parts. The first part is more introductory and talks about what the Internet of Things is, how it is related to Cloud computing and what are the basic principles behind sensors and actuators that are found in most common IoT projects.

The second part introduces you to the microcontrollers and the Arduino platform: what it is, how you can program your Arduino to sense the environment and make it communicate with the Internet using wired, wireless networks and even your Android-based phone. Projects described in the chapters include programming and using a microcontroller out of the board, using timers, threads and encryption libraries for your IoT projects, extending Arduino, testing your code using an Arduino emulator, interfacing with analog and digital sensors, log air quality data and make tweets through your Android phone, control a relay switch by texting your phone, Internetize your Arduino through various ways and even send your sensor data to your own Cloud-based application.

The third part focuses on the use of existing Cloud applications for managing your sensor data. The Cosm and Nimbits are presented as the most common platforms for storing and visualizing readings from your Arduino board. This part also demonstrates how to reprogram your Arduino remotely using your own Cloud-based web application. The final Chapter includes ideas of what you can connect to your Arduino and communicate it with the Internet.

How to Use The Book

The main purpose of this book is to introduce you to the concept of managing sensor data on the Cloud. It teaches you the basics of Arduino, how to enable it to communicate with the Internet and web applications on the Cloud. Each chapter serves a specific purpose and starts with a short introduction of the concept it presents. You are advised to read the chapters in the given order, especially those in the first two parts of the book, so that you are smoothly introduced into the concepts of Cloud computing and the Internet of Things. You will be able to familiarize yourself with the Arduino platform and then start building projects that communicate sensors and actuators with the Internet.

Each project presented starts with describing the basic functionality it provides and then with listing the essential hardware components. Circuit schematics and components connections are also illustrated in appropriate figures.

In case you find hard to read connections in figures, you can always look at the **figure repository** at http://www.buildinginternetofthings.com that contains full color images at higher resolutions. The software code is presented and explained. The code is initially listed the way it should be programmed, and then explained thoroughly line by line so that you can understand what it does and how it does it. Therefore you are advised to read each project from the beginning, understand what it is expected to do, read about the components you need to build the circuit, and then move to the code listings. Read the complete code once. No matter how experienced you are with coding and the Arduino, you will find some parts (or the whole code) unfamiliar. Then move to the respective code review sections and you will find all the answers to your questions. Collect the essential components and build the circuit. Program your Arduino board (and creating Java and Android applications in some projects) and finally see the expected outcome yourself.

When coming to connecting your Arduino to the Internet, you will notice that there are several ways to do so: using your Android phone or your computer as intermediate gateways or by directly connecting it through a wired (Ethernet) or wireless (WiFi) module. Each way has its own pros and cons and cost. You are welcomed to use whatever way suits best your needs and meets your available hardware. The book contains code samples and instructions that cover all ways.

Knowledge Required

This book assumes that you are already familiar with the general principles of software programming and also familiar with programming in C/C++ and Java. Especially for the latter, it is also assumed that you are familiar with using and creating projects in Eclipse Java IDE.

If you are a total beginner in programming you can still use the book but first you are advised to study any of the following introductory books in Arduino, Java and Android programming:

Java 7 for Absolute Beginners, by Jay Bryant

Beginning Android, by Mark Murphy

Chapters 6 and 8 also describe projects that include Android code. So you also need to be familiar with Android programming and how to set up the Eclipse IDE for Android development.

In addition, you need to be familiar with the basics of electronics and how you can use a breadboard to connect various electronic components together.

What You Will Need

To make your Arduino interact with its environment and communicate over the Internet you need both appropriate hardware and software.

Hardware
For all the projects covered in this book you need at least one Arduino board. You can get either an official Arduino board (see Chapter 4 for more details) or an Arduino-compatible clone (like in Figure 1). In order to program the board you will need a computer (Windows, Mac or Linux), and a USB cable (type depends on the board you have).

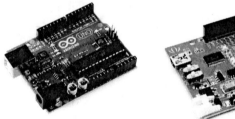

Figure 1. On the left: An official Arduino Uno board (image courtesy of Sparkfun). On the right: The Seeeduino v2.21 compatible with the Arduino Uno (image courtesy of Seeedstudio)

In addition, the various projects covered in this book require various electronic components (such as resistors and capacitors), sensors (like temperature and humidity sensors) and actuators (mostly a relay switch).

For the communication with an Android phone you will need either a capable board that connects directly with your phone over USB or a Bluetooth enabled board (more details for both ways on Chapter 6).

For communicating your Arduino directly with the Internet you can use an Ethernet-enabled Arduino board or a WiFi shield (more details in Chapter 7).

All essential parts and the way to connect them together are listed in the project description of each chapter. You are also advised to use a breadboard (see Figure 2) and jumper wires (male-to-male) (see Figure 3) in order to easily connect the various components with each other and with the Arduino.

Places where you can buy online all the components featured in the projects of this book are the following:

- Sparkfun electronics, **http://www.sparkfun.com**
- Seeedstudio electronics, **http://www.seeedstudio.com**

Both electronic stores ship worldwide. Of course you are free to search online for alternative stores that provide the materials you will need (EBay can also another good source).

Figure 2. A breadboard (image courtesy of Sparkfun)

Figure 3. A male-to-male jumper wire. It can be easily connected to the breadboard and the Arduino pins (image courtesy of Sparkfun).

Software

In order to program your Arduino board you will need the Arduino development environment (or IDE) that you can freely download from the Arduino web site (**http://arduino.cc/en/Main/Software**) and use it on your Windows, Mac or Linux computer. Detailed instructions on how to setup and use the IDE are provided in Chapter 4.

Some of the projects in the book also teach you how to create standalone Java applications and Android-based applications that communicate with your Arduino. In those cases you will need to have installed and be familiar with the Eclipse Java Development Environment (**http://www.eclipse.org**).

The Code

As of writing this book, Arduino has released the v1.0.1 that introduces some new features and changes in the way code and especially libraries are written (compared to previous IDE version like 022). The Arduino code (aka sketches) provided in the book can be run and compiled under v1.0.1. Many of the sketches rely on external libraries that were built for previous Arduino versions.

All libraries have been ported to v1.0 and are included in the source code that can be downloaded from the book's online **code repository** found at: **http://www.buildinginternetofthings.com**. The code repository will be frequently updated reflecting any future changes in the Arduino IDE, the IoT platforms presented in the book and potential improvements.

Note on Version 1.1

This is a second release of the book. It contains updates, corrections and improvements. The updates reflect the recent changes of Pachube platform (now Cosm), the release of a new Arduino board (Arduino Leonardo) and additional references to new IoT Cloud platforms.

Contacting the Author

You can contact me for questions, comments, corrections and suggestions at ch.doukas@gmail.com. You can also follow my Twitter account @BuildingIoT for instant updates and news about the IoT.

I hope you will enjoy reading this book as much as I did writing it. Much more I hope the projects described will guide you in designing and building your own projects!

Charalampos Doukas
Athens, June 2012

PART I

INTRODUCTION TO THE INTERNET OF THINGS

"When we talk about an Internet of things, it's not just putting RFID tags on some dumb thing so we smart people know where that dumb thing is. It's about embedding' intelligence so things become smarter and do more than they were proposed to do."

Nicholas Negroponte

1 THE INTERNET OF THINGS

The Internet has initially started as the "Internet of Computers", a global network enabling services that now include the World Wide Web (WWW), File Transfer Protocol (FTP) and others allowing computers and hence users to communicate with each other and exchange information. In the recent years, the device processing power and storage capacity are increasing while at the same time technology is making devices pervasive, mobile and wearable. In addition, networking technologies are evolving and communication electronic systems are becoming smaller and cheaper. Devices are increasingly fitted with sensors and actuators, creating environments where the former are connected to various networks. Devices can sense, compute, act and thus intelligently become parts of the so-called 'Internet of Things'.

There are several definitions for the Internet of Things (IoT) that also explain what are the main functionalities of it and what we should expect from when connecting 'Things' with each other and with the Internet. Some people suggest, that the "Internet of Things can be seen as a potentially integrated part of the 'Future Internet'. Wikipedia defines IoT as:

"A part of a dynamic global network infrastructure with self-configuring capabilities based on open and interoperable communication protocols where physical and virtual 'Things' interact with each other. These 'Things' have specific identities, physical attributes, virtual 'personalities' and use intelligent interfaces. They are able to interact and communicate among themselves and with the environment by exchanging data and information 'sensed' about the environment, while reacting autonomously to the 'real/physical world' events and influencing them by running processes that trigger actions and create services with or without our direct intervention. Interfaces in the form of services facilitate interactions with these 'smart things' over the Internet, query and change their state and any information associated with them."

To people like hobbyists, electronic enthusiasts or sensor researchers the IoT is new opportunity and at the same time a new challenge for managing the data we acquire from our embedded electronics projects and controlling their outputs.

Imaging having a small device at the size of a matchbox, that can senses temperature, humidity and light conditions of your room, and can report them directly to a web-based service. The readings by the sensors can be accessed only by you through your favorite browser, by your mobile phone and by other devices in you place, like the central heating /air conditioning system or the indoor lights control system. The latter can adjust the heat and lighting inside your place automatically, making sure you have always the most preferable conditions as defined in the web-based service by you.

Now imagine that you don't have to build everything from scratch in order to develop and deploy such a system. Imagine also that you do not have to worry about data management; how is the data stored on the web, what kind of web server and web application technology you have to use for your service, how to secure your data and implement various authentication and encryption mechanisms! You do not even have to worry about learning how to develop mobile applications that talk to your service! The web-based services that acquire and manage your sensor data, the mobile application, the communication interfaces and information exchange protocols are already available for you to exploit and explore them! What are they? They are applications and features of existing platforms that provide users with Internet of Things functionality. Where are they? We will explore a few of them and ways to use them in the following chapters of this book. Firstly, let's take a closer look at the concepts of IoT.

The Basic Concepts

Let us briefly discuss the main concepts and essential components that are mostly used in order to describe the world of the Internet of Things.

When we are referring to 'Things', we talk about devices and everyday objects, from small ones (like wrist watches and medical sensors) to really big ones (like robots, cars and buildings). All such contain devices that interact with users by generating and retrieving information about and from their environment (see Figure 1-1). They also contain hardware that allows them to control outputs (like relay switches or digital ports).

No matter what definition of the Internet of Things you may find, the main concept behind every IoT technology and implementation is the same: devices are integrated with the virtual world of the Internet and interact with it by tracking, sensing, and monitoring objects and their environment. Users and developers add components, give them sensing and networking capabilities, program them to perform the aforementioned tasks and build Web applications that interacting with the devices.

The features of a device that can act as a member of an IoT network can be summarized into the following:

- Collect and transmit data: The device can sense the environment (e.g., your home or your body) and collect information related to it (e.g., temperature and lighting conditions) and transmit it to a different device (can be your mobile phone or your laptop) or to the Internet.

- Actuate devices based on triggers: It can be programmed to actuate other devices (e.g., turn on the lights or turn off the heating) based on conditions set by you. For instance, you can program the device to turn on the lights when it gets dark in your room.

- Receive information: One unique characteristic for IoT devices is that they can also receive information from the network they belong to (i.e. other devices) or through the Internet (e.g., information from you like new triggers, new status of operation and in some cases new functionality).

- Communication assistance: IoT devices that are members of a device network can also assist in communication (i.e. data forwarding) between other nodes of the same network. Think of them as messengers for devices (nodes) that are not very close to an endpoint (e.g., your router) in order to get direct information from.

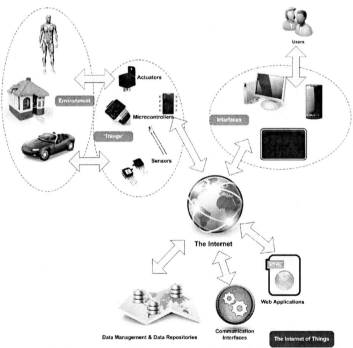

Figure 1-1. An illustration of the 'Internet of Things'. 'Things' consisting of various sensors and actuators interact with environment and the Internet allowing users to manage them and their data over various interfaces.

Interaction with the Internet

What makes IoT devices different than ordinary sensor devices is basically the ability to communicate (most usually) directly or indirectly to the Internet. So what are the main reasons a device would need to communicate with an Internet service? What kind of service would that be and what kind of features would it have? Firstly, sensors generate a lot of data that needs somehow to be managed. Usually, embedded memory is quite limited so people utilize alternative solutions like storing data on memory cards, or in computers in cases sensors are directly connected to them. Since sensors can be integrated to devices with further networking capabilities, why not to store the information online? Through this way we can solve the limited storage issue and at the same time we can access the data anywhere, anytime using appropriate web applications. Figure 1-2 illustrates the main features of 'Things' and their interaction with Internet services.

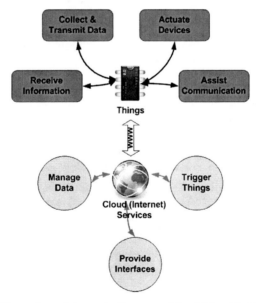

Figure 1-2. Illustration of the main features of the 'Things' in IoT networks and the respective Internet services.

In addition, most of these web applications provide interfaces for data exchange between other applications and most usefully, with mobile applications. So it is possible that your iPhone or Android mobile can talk directly to your IoT device!

We can also use the web platforms to send data back to the devices. The data can be triggers, instructions for activating actuators and switches, or simply information from the Internet, like a weather channel or a Twitter account that can displayed through an interface like an LCD screen.

The web-based platforms that enable the aforementioned functionality for data management and information exchange are based on Cloud computing. More information on Cloud computing platforms will be presented in Chapter 3 of this book.

In order for things to talk to each other and to the Internet, they need to have specific abilities and serve more specific functions. The following section discuses what are the components that are included in IoT devices that enable them to sense the environment, act and communicate.

Major Components of IoT Devices

Let's discuss what are the main components of an IoT device. Namely, such components are the main control units (the 'brains' of the devices), the sensors that collect information – signals from the environment, the communication modules and the power sources.

Control Units

Usually in the world of IoT, devices utilize a microcontroller as the main control unit responsible for the aforementioned operations. A microcontroller can be considered as a small computer on a single integrated circuit (often abbreviated as IC) containing a processor core, memory, and programmable input and output peripherals. Figure 1-3 presents a very popular microcontroller chip. The peripherals are controlled through several general-purpose input/output pins, known as GPIO pins. GPIO pins are configured to either input (read) data or output data. When they are configured to an input state, they are used to read information from sensors and accept external signals (e.g., button events). Configured to the output state, GPIO pins can control external devices such as LEDs, motors, relay switches, etc. In addition, these pins are also used by the controller to communicate with devices, such as modems and other communication modules.

Other important parts of the microcontrollers are the analog to digital converters (ADC) that are used to convert incoming analog (signal) data into a form that the processor can recognize. In addition to the converters, many such control units include a variety of timers as well.

We will discuss more on microcontrollers and their components in the following chapters.

Additional modules like Pulse Width Modulation (PWM) modules are also included in the microcontroller and enable it to process and control more advanced devices like power converters and, motors without using lots of resources in generating pulse signals programmatically.

Figure 1-3. The Atmega328 chip from Atmel. One of the most widely used microcontrollers in embedded projects. Each pin has a specific function; receive analog signals, communicate through serial interface with other components, receive input voltage, generate pulses or digital output, etc. (image courtesy of Sparkfun).

Sensors

Sensors are devices that can measure a physical quantity (like temperature, humidity, etc.) and convert it into a signal, which can be read and interpreted by the microcontroller unit. They are the devices that are most likely attached to the input pins of the microcontroller in our embedded projects. Generally, most sensors fall into two categories; analog and digital sensors.

An analog sensor, such as a thermistor (which actually is a resistor changing its resistance based on temperature used in digital thermometers), or a humidity sensor, is connected into circuits so that it generates a specific voltage range, usually between 0 volts to 5 volts. This output goes to the analog input pin of the microcontroller unit (see Figure 1-4). The latter uses the appropriate analog-to-digital (A/D) circuit to convert voltage into a numeric value that we can later read, process and send to the Internet. Figure 1-4 shows how an analog sensor can be connected to a microcontroller device. All analog sensors usually come with 3 connection pins, one for receiving voltage (Vcc), one for ground connection (GND) and the output voltage pin (OUT).

Digital sensors generate what is called a Discrete Signal. Such a signal has different (usually binary only) states, like the logic gate states 0 and 1, which are represented by high and low current. For example, consider a push button switch (which by the way is one of the simplest forms of digital sensors): it has two discrete values. It is on, or it is off. Some more complicated sensors come also with a digital interface. This means that instead of generating a signal representing the voltage change between 0-5 based on their measurement, they actually generate a stream of bits (0 and 1 states).

More information on how sensors work is provided in chapter 2 of this book.

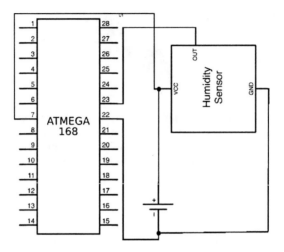

Figure 1-4. This is how a humidity sensor connects to the Atmega168 microcontroller. Vcc receives operating voltage and GND is connected to the ground of the power source. OUT is the pin that outputs the sensed signal and connects to microcontroller pin 23 (analog port).

Communication Modules

The communication modules are parts of the device, which are responsible for the communication with the rest of the IoT platform. They provide connectivity according to the wireless or wired communication protocol they are designed for (such protocols are presented in the rest of this chapter). They usually consist of embedded electronic modules that implement the communication (i.e. transform the information received in bits and bytes to radio waves or signals that are transferred by wire respectively). Wireless modules usually consist also of an antenna for achieving maximum range, either external or internal for optimizing the size of the device.

Most likely the communication between IoT devices and the Internet is performed in two ways: a) there is an internet-enabled intermediate node acting as a gateway, b) the IoT device has direct communication to the Internet. In the first case, the device is commonly connected to the computer and sends data to it using e.g., a USB port. The computer receives the data and using appropriate software it forwards it to the Internet. In the second case, things are much simpler and devices can function and communicate more autonomously.

In either case, communication can be made according to the wireless technology used. Most modules that support wireless (like WiFi, Bluetooth and ZigBee) and wired technologies (like the Ethernet) also support the TCP/IP protocol. The TCP/IP protocol (for those unfamiliar with it) is the way computers, mobile phones and internet-enabled devices communicate with each other and to the Internet. It defines all communication essentials (e.g., how devices are identified by each other, using IP addresses; how information is split into small packets and transmitted; etc.). It also takes care of all connection setup issues and ensures information is properly delivered by retransmitting data whenever this is considered necessary.

What about inter-device communication? The communication between the main control units and the various communication modules is performed in most cases (and in the case of our projects) using the serial protocol. Serial is a device communication protocol that is standard on almost every PC and electronic device. The concept of serial protocol is simple. Each device has two ports (or connectors), one for transmitting data, usually annotated as Tx, and one for receiving data, usually annotated as Rx. Also a ground and power connection are required and are provided by the power source. Each serial port sends and receives bytes of information one bit at a time. So, in order to communicate, both control units (i.e. microcontrollers) and communication modules share a pair of Rx/Tx ports, as in Figure 1-5.

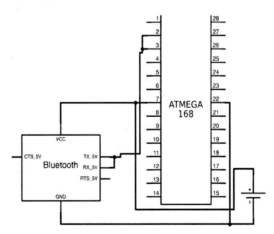

Figure 1-5. This is how a Bluetooth communication module connects to the Atmega168 microcontroller. Vcc receives operating voltage and GND is connected to the ground of the power source. Tx and Rx pins are the pins that send and receive serial data and connect to microcontroller pins 2 and 3 (Rx and Tx respectively).

Although this is slower than parallel communication, which allows the transmission of an entire byte at once, it is simpler and you can use it over longer distances. Because serial is asynchronous, the port can transmit data on one line while receiving data on another.

A brief presentation of communication enabling technologies for IoT devices follows in the "Communication Technologies" section.

Power Sources

Every electronic device needs electric power to properly function. The electric power is the result of the potential difference between two points, otherwise known as voltage, and the electric current flowing through a circuit. In most electronic devices and also in the projects we are dealing with in this book, the two points are referred as the voltage input point (usually notated as Vin or Vcc) and the ground (usually notated as GND).

In small devices the current is usually produced by sources like batteries, thermocouples and solar cells. Batteries are electrochemical components that convert chemical power to electrical generating direct current (DC).

Wearable and mobile devices are mostly powered by small lightweight batteries that can also be recharged (like in Figure 1-6) for longer life duration. By utilizing simple circuits and appropriate programming of microcontrollers, devices can be aware of their battery status and alert users when they need to be recharged or replaced.

Figure 1-6. Probably the smallest rechargeable consumer battery. Weighs only 330mg and operates at 3.6V (image courtesy of PowerStream)

When selecting the appropriate battery source for your own projects, you look for the battery's operational voltage and capacity. The latter is associated with the total energy stored within the battery. Capacity in general is measured in watt per hours (Wh), kilowatt per hours (kWh) or ampere per hours (Ahr). For sensors and other components of IoT projects you will build, the most common measure of battery capacity is Ah.

To give you an example about capacity and its actual meaning, if your IoT device needs 50mA to operate and your battery provides 50mAH, the latter will power your device adequately for about 1 hour before needing recharging.

Communication Technologies

'Things' need to talk to each other and also talk to the Internet in order to exchange sensor outputs, triggers, status messages etc. In order to do so, devices need to integrate a wireless (preferably) or a wired communication system. The major communication technologies that can be utilized by such devices are summarized in this section. In addition to the brief description of the technologies, samples of electronic modules that enable the respective communication are also presented. The modules selected are characterized by the special interfaces they have for connecting directly to microcontrollers and platforms like the 'Arduino' open-source electronics prototyping platform.

RFID

The Radio Frequency Identification (RFID) technology has been initially introduced for identifying and tracking objects with the help of small electronic chips, called tags. It is the most common technology behind asset tracking (that tells you where your mail parcel is before arriving its destination) and identifying objects (e.g., in automatic toll collection).

RFID tags are categorized into passive, active and battery assisted passive (BAP). The passive tags do not have a power source (battery) and thus cannot transmit and information on their own. They are powered-activated by the RFID reader and transmit only a small amount of information (usually an identification number (ID) of the tag). Active tags on the contrary have their own battery and can broadcast data continuously. A BAP tag can be considered as a hybrid: it carries a battery but only transmits information in presence of an RFID reader. The battery helps them to transmit their signal in longer distance than the passive tags (restricted to a few cm).

Figure 1-7. RFID module and its antenna. It can directly communicate with the Arduino using the Serial protocol. (image courtesy of Seeedstudio).

Most RFID tags consist of an integrated circuit for storing and processing the information, the essential radio frequency (RF) components for transmitting the information wirelessly and an antenna.

RFID has been initially characterized as the enabling communication power for the Internet of Things, due to its low cost, high mobility and efficiency in identifying devices and objects. Figure 1-7 illustrates an RFID module and its antenna that can directly communicate with a microcontroller.

Despite RFID being very popular for device identification and some information exchange (e.g., RFID-based temperature sensors) it cannot alone support the creation of IoT networks since it cannot provide any direct or indirect (e.g., through a gateway) communication to the Internet. The device proximity is also another drawback.

Bluetooth

You are using the Bluetooth technology to connect your wireless headset with your mobile phone, or to transfer photos from the latter to your computer. Bluetooth is a technology standard for exchanging data over short distances (using short wavelength radio transmissions in the ISM band from 2400-2480 MHz) from fixed and mobile devices, creating personal area networks. Bluetooth has been one of the first wireless communication protocols designed with lower power consumption for replacing short-range wired communications (in computer peripherals, mobile phone accessories, etc.). One important feature of Bluetooth is that devices can discover and communicate with each other without the need to be in visual line of sight (like in infrared communication), which is very important when using Bluetooth as a network technology for sensor systems deployment.

Bluetooth is named after the 10th century Danish King Harald Blåtand, which translates as Harold Bluetooth in English!

It is commonly used for connecting small devices with each other, due to the fact that it can support automatically the creation of peer networks (i.e. networks of devices that exchange and forward information) and provides communication functionality with low power consumption. The latter is very important for the case of IoT since many of the devices that one would like to interconnect to the IoT (sensors, actuators, etc.) have limited power resources. However on major drawback of Bluetooth is that it cannot provide direct connectivity to the Internet. One has to provide an intermediate node, e.g., a PC that will act as a gateway to the outer world. You can see how a Bluetooth communication module looks like in Figure 1-8.

Figure 1-8. Bluetooth communication module on a breakout board for easily interfacing with devices like the Arduino. The module is so small it fits in your mobile phone, your hands free speaker and a USB dongle (image courtesy of Sparkfun).

ZigBee

ZigBee is one of the latest and most advanced wireless technologies being widely integrated into home automation & smart devices worldwide. It has been specifically developed as an open global standard to address the unique needs of low-cost, low-power wireless networks for communication between devices (also known as machine-to-machine or M2M networks). The ZigBee standard operates in unlicensed bands including 2.4 GHz, 900 MHz and 868 MHz

A ZigBee module (see Figure 1-9) can have a low consumption of 50mA. For instance, a rechargeable battery of 850mAH can provide about 17 hours of continuous operation for such module. Maximum data rate is about 250kbps and communication range can vary from 100m to 1km (maximum) depending on the output power (in small projects we usually use 1mW modules that provide maximum of 100m range).

Figure 1-9. ZigBee communication module in a pluggable form known as XBee series modules (image courtesy of Sparkfun).

Compared to Bluetooth, ZigBee provides better power efficiency, and higher range, making it thus a better wireless technology to consider for your IoT network. It also allows the automatic creation of peer networks and as a matter of fact it is consider being more extensible in this context.

However it still requires a gateway with Internet connectivity (e.g., a laptop) that will forward information from the Internet to the ZigBee network and vice versa.

WiFi

WiFi, also known as the IEEE 802.11x standard, is the most common way to connect devices wirelessly to the Internet. Your laptop, smartphone and Tablet PC are equipped with WiFi interfaces and talk to your wireless router and provide you this way access to the Internet.

The commercially available WiFi modules (like the one in Figure 1-10) can be directly integrated to an IoT device and provide instant connectivity. The major advantage over the other wireless technologies is the fact that WiFi networks are very easy to establish and thus IoT devices with WiFi modules can have direct connection to the Internet. One drawback is the fact that this technology (which was at no means designed for IoT networks) is more power demanding than the others.

Figure 1-10. WiFi communication module in a pluggable form known as XBee series modules (image courtesy of Seeedstudio).

RF Links

Another option to connect devices and make them talk is utilize simple radio frequency (RF) interfaces. The latter are quite cheap and small (ideal when size matters) and can provide communication range between 100m and 1km (depending on the transmission power and the antenna used).

RF communication modules (see Figure 1-11) are connected to microcontroller and devices via through serial ports as the rest of the modules we have examined. However they do not provide any implementation of the TCP/IP communication protocol (or any other protocol). This means that if you want your devices to communicate you have to create your own protocol for establishing communication, identifying devices with each other and make sure all the information you have transmitted is delivered. Data rates are quite low (up to 1Mpbs) and you also need an Internet-enabled gateway that will provide access to your devices for making a complete IoT network.

Figure 1-11. An RF transceiver module (it can be used both as a receiver and transmitter). It has low power consumption, can support data rates up to 1Mbps and can be easily interfaced to your Arduino (image courtesy of Sparkfun).

Cellular Networks: The Mobile Internet

The 'Mobile Internet' refers usually to access to the Internet from a mobile device, such as a smartphone or laptop through a mobile broadband network. The mobile broadband network is based on cellular communication, same technology that is used in our mobile phones for serving our calls and text messages. It can provide direct Internet connectivity to a variety of data rates. Various network standards have existed for serving mobile Internet. GPRS, 3G, WiMax, and LTE (one of the 4G technologies) are a few to name. Depending on the standard and the available network coverage, connection speeds can go from 80Kbps (GPRS) to a few Mbps (3G and 4G).

Due to the complexity of the communication protocol and the information coding, in addition to high power requirements in cases where reception signal is low, the battery consumption of mobile Internet – enabled devices is an issue. Think of how fast the battery of your cell phone is drained when you browse the Internet. Still it is a good option for connecting devices directly to the Internet, since small GPRS modules for the Arduino are available (see Figure 1-12) and connectivity does not require further infrastructure (e.g., Internet connected laptop like in case of ZigBee or Bluetooth).

Figure 1-12. GPRS shield for your Arduino on the left. The GPRS module on the right. The small size of it allows it to fit easily in mobile phones, Tablet PCs and sensor devices (image courtesy of Sparkfun).

Wired Communication

Devices can definitely talk to each other and to the Internet using wired infrastructures, given the appropriate network interface. What is the best way? Since the very beginning of the Internet era to nowadays, your desktop computer has at least one Ethernet port.

The Ethernet protocol is very well established in computer communications. It does not require as much power as the wireless communications do, it can achieve very high data rates and most importantly it is so common in computer communications that all you need to do is plug an Ethernet cable to its device and get online. No worries about signal availability. It does not provide any mobility, but its advantages were the main reason it was one of the first networking technologies adopted in microcontroller platforms like the Arduino. Figure 1-13 presents an Ethernet module that can be directly plugged into the Arduino and an Ethernet-enabled Arduino board.

Figure 1-13. On the left: An Ethernet module. Can be directly connected to any sensor device and provide wired network functionality. On the right: An Arduino with embedded Ethernet interface (image courtesy of Arduino).

Which One Is the Best?

When coming to the point where we need to decide what is the most appropriate communication technology for your IoT network, there are several things we need to consider; mobility, network range, power consumption, size and cost are the most important to name. Table 1 – 1 provides an overview of the presented communication technologies comparing their features.

In case your devices will be at a fixed position, then an Ethernet module is one of the best options for providing communication at high-speed ranges with low power consumption. In a different case where mobility is a must (or there is no wired infrastructure available), we can select between WiFi, cellular and ZigBee. RF requires much more effort for building a communication protocol. ZigBee requires a gateway for providing connection to the Internet, but otherwise it is a very good solution, providing low power consumption and good coverage range. WiFi can provide direct access in case a WiFi infrastructure is available, but consumes much more power since it was originally designed for devices like laptops with power resources not for small IoT devices. When the only network infrastructure available is a mobile network, cellular data communication is the only option.

Table 1-1. Overview of the Wireless Communication Technologies.

Technology	Data rate	Range	Frequency
WiFi	54 Mbps	150 m	5 GHz
Bluetooth	721 Kbps	10 - 150 m	2.4 GHz ISM
RF – links	1 Mbps	50 – 100 m	2.4GHz ISM
IEEE 802.15.4, ZigBee	250 kbps, 20 kbps, 40 kbps	100 - 300 m	2.4 GHz ISM, 868 MHz, 915MHz ISM
Cellular 3G	14.4 Mbps/ 5.8 Mbps	m - km	800 MHz, 1900 MHz
Wired (Ethernet)	100 Mbps – 1 Gbps	m - km	-

Current Status and the Near Future

What has been the progress of the IoT since the initial description of the concept back in 1999? The evolution of the IoT in our daily lives depends mainly on two things; on the technical progress of microcontrollers and embedded devices and on the evolution of wireless interfaces in terms of communication and power efficiency.

Regarding the services for the IoT networks the current status involves several platforms that are already providing functionalities like hosting of data repositories, visualization of data, communication with the devices through web interfaces, remote triggering of events, etc. All these platforms and their services are specially designed and implemented for enabling the communication with IoT devices. This means that their functions are implemented using open and lightweight Internet web protocols that allow easy and direct communication, offering at the same time implementations in many programming languages and environments.

Examples of IoT services include the ones that will be presented in this book, like the Cosm data service and the Nimbits data logger and many more like the ThingSpeak and the iDigi Device Cloud. These services allow users to manage and visualize their sensor data either free or at low cost. Additional services like the recently discontinued Google Health or the Microsoft HealthVault collaborate with device vendors for managing device information on the web directly. Internet-enabled blood pressure monitors, glucose meters, pedometers, and similar health monitoring devices are available in the market enabling the user's to manage their health data from their web accounts.

The near future vision: It is already feasible to connect microcontroller platforms, like the Arduino, directly to IoT services and having them report sensor data and response to triggers set by users remotely through web applications. Embedded devices are already small enough to be even wearable, and advances in communications and power resources enable wireless Internet connectivity to all kinds of devices that we currently use. Imagine being able to control your home or office automation the way you view emails today; either from your mobile, your laptop or any place on the world with an Internet connection. Most importantly and much more interestingly, imagine that at the same time your automation devices will be able to retrieve important information for their function (e.g., weather updates) directly from the Internet make decisions, provide automations and require less input and interaction from you.

Explore the examples of this book and become ready to fully 'Internetize' your sensors and actuators by building your own IoT network with your devices!

Summary

This chapter has introduced you to the basic concepts of the Internet of Things. You have been provided with examples of IoT devices along with information about the major components (like microcontrollers and sensors). A brief presentation and comparison between the main communication technologies (like WiFi, Bluetooth, RF, Cellular, ZigBee and Ethernet) can give you ideas about how the IoT devices can interact with the Internet. The last section of this chapter has provided you information about the current status of the Internet of Things evolution and given you ideas about the near future.

The next chapter will explain you how sensors work and what kind of sensors you can utilize in your embedded projects for building your own IoT platforms.

2 THE BASIC OF SENSORS AND ACTUATORS

Building your own 'Internet of Things' requires from you to decide and define what you will have connected to the IoT and what kind of functions it will perform for you. This brings us to the second chapter of this book that presents the basics of sensors and actuators. It will help you understand what you can connect to your Internet-enabled devices, what you can sense from your environment through them and how you can control your environment remotely.

We will start by explaining what sensors are and how they work. Part of this presentation also includes some brief explanation of sampling theory and conversion of analog signals to digital information. Additionally, there is a list with the most usual sensors and actuators you could use for your own projects.

Introduction

Let's start by discussing what a sensor is: A sensor is a device that can measure a physical phenomenon (like temperature, radiation, humidity, etc.) or detect chemical concentration (e.g., smoke) and quantify the latter, in other words, provide a measurable representation of the phenomenon on a specific scale or range (e.g., a voltage range). Sensors are usually attached to the input pins of the microcontroller in our embedded projects and are used by a great variety of devices in our daily lives (e.g., laptops, mobile phones, home automation, automobiles, etc.). Generally, most sensors fall into two categories; analog and digital sensors.

Analog Sensors

To understand what an analog sensor is and how it functions, it is necessary first to make a short introduction about resistors (in case you are not familiar with them). The reason we need to do this is because most analog sensors when attached to a circuit operate like resistors. A resistor is a portion of an electronic device that lowers the voltage between its two terminals. The voltage drop is intended to oppose, or "resist," an electrical current. The drop in voltage is inversely proportional to the amount of current that is being opposed. Resistors are composed of several elements and can be in many different forms. They are part of most types of circuits.

An analog sensor, such as a thermistor (resistor changing its resistance based on temperature), or a humidity sensor, is wired into a circuit in a way that it will have an output with a specific voltage range, usually between 0 volts to 5 volts. This is what we call an analog signal, which is continuous in both time and amplitude (see Figure 2-1). The output goes to the analog input pin of the microcontroller unit that uses the appropriate analog-to-digital (A/D) circuit to convert voltage into a numeric value that we can later read, process and send to the Internet. An example of how an analog sensor can be connected to a microcontroller device has been demonstrated in Chapter 1 (see Figure 1-4). All analog sensors usually come with 3 connection pins, one for receiving voltage (Vcc), one for ground connection (GND) and the output voltage pin (OUT).

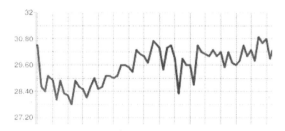

Figure 2-1. An analog signal as produced by an analog temperature sensor measuring temperature in my home. The signal is continuous in both time and amplitude.

Digital Sensors

As its name implies, digital sensors produce a discrete output signal or voltage that is a digital representation of the quantity being measured. Digital sensors produce a binary output signal in the form of a logic "1" or a logic "0", ("ON" or "OFF"). In voltage, usually '1' is represented by 5V and '0' by 0V. This means then that a digital signal only produces non-continuous values (what we call discrete values) which may be output as a single "bit" (serial transmission) or by combining the bits to produce a single "byte" output (parallel transmission).

Compared to analog signals, digital signals are much more accurate (less tolerant to noise compared to analog ones) and can be both measured and "sampled" at very high clock speeds. The accuracy of the digital signal depends on the number of bits used to represent the measured quantity. Figure 2-2 illustrates an example of a digital signal.

The simplest version of a digital sensor is a button switch, as it has two states, on and off defined by user interaction. It generates either high voltage (5V) in the connected digital input pin of the microcontroller, or low voltage (0V) respectively. To read its input you simply connect a circuit to the digital input port of a microcontroller (you might add a pull-up resistor as explained hereafter) and check wherever a high or low input is detected when closing or opening the circuit. More advanced versions of digital sensors, are those equipped with integrated circuits - processors that convert the analog signal of the physical quantity to bits and bytes. The latter are transferred through an output pin to the microcontroller. Interpreting the incoming bits (indicated as low and high states) requires some additional coding and the knowledge of the communication protocol (e.g., how to assemble bytes, which bytes contain useful information, etc.) provided by the vendor. Despite being a bit harder to use, such digital sensors can provide advanced information about the measured phenomenon and can also combine measurements (e.g., temperature and humidity) in one sensor device.

To find out more about sensors and how Arduino can communicate with digital sensors check Chapter 5.

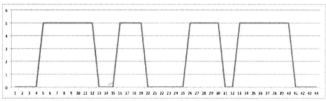

Figure 2-2. An example of a digital signal.

Pull-Up/Down resistors and Sensors

Do you connect sensors directly to the pins of your microcontroller? In theory yes, but in practice, as you will also notice in many electronic project examples, you usually use small resistors, called pull-up or pull-down resistors.

Pull-up resistors are resistors that make sure that logic system inputs (i.e. input ports of a microcontroller) remain at the correct levels even when devices like sensors are removed from them. They are named like this because they "pull" the voltage of the wire up to a pre-calculated high voltage. The pull-up resistor is left intentionally weak, however if another device pulls the voltage of the wire to another voltage, the pull-up resistor will not affect the circuit. The primary function of a pull-up resistor is to prevent excessive current from flowing through the circuit and they are therefore preferred in many circuits.

The Arduino microcontroller environment comes with a built-in pull-up resistor. This means you can directly connect sensors on the analog/digital ports, however you should keep in mind the usability of a pull-up resistor.

As an alternative to pull-up resistors you can use (or notice being used in sensor circuits) a pull-down resistor. Just like the pull-up resistor, it is used to limit the current that can flow between Vcc and ground. For such purposes you can use either a 10K Ohm or a 47K Ohm resistor in your project. The exact value doesn't really matter, as long as it is high enough to prevent too much current from flowing towards the input port of your microcontroller.

Let's see a simple example for the usage of a pull-up resistor. Consider a circuit with a button switch and a microcontroller as in Figure 2-3. Your goal is to identify through the microcontroller whenever the switch is closed or opened. Let's connect on pin of the switch to the Ground and the other to a 5V source through a 10K ohm resistor (your pull-up resistor). When the switch is open, current flows to the digital port of the microcontroller indicating a high state. When closing the switch, current will flow from the 5V source through the resistor to the Ground indicating a low state at the microcontroller. If no pull-up resistor was present, a short-circuit between the 5V source and the Ground would occur.

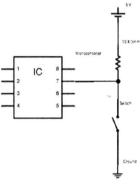

Figure 2-3. Connecting a pull-up resistor to a digital sensor (button switch) circuit.

The alternative way of connecting the circuit and using a pull-down resistor is displayed in Figure 2-4.

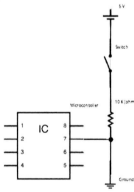

Figure 2-4. Connecting a pull-down resistor to a digital sensor (button switch) circuit.

28

A brief Introduction to sampling theory

What is an analog signal? Basically everything we perceive by our senses in the real world: sound, colors, smells, pretty much anything! We are 'built' to receive such signals and transform them into something we can understand (guess what the microcontroller is in this case), but what about computers? For such digital devices we need to convert the analog signals to digital form. A digital signal is discrete in both time and amplitude (as opposed to the analog signal that is continuous). In order to convert an analog signal from our world to a digital signal so that computers and microcontroller can perceive it, we use a technique called 'sampling'. Sampling refers to dividing the original signal into many small time intervals, called samples and measure the value of it for each sample. Then the respective signal value of each sample is converted to binary format. This is usually performed by the microcontroller using special Analog-to-Digital (A/D) conversion circuits (see Figure 2-5).

The sampling rate (i.e. how often we divide the original signal into samples) is also called sampling frequency. How many samples are necessary to make sure we preserve the information contained in the signal? If the signal contains high frequency components (like human speech does), we need to sample at a higher rate to avoid losing information that is in the signal. In general, to preserve the full information in the signal, it is necessary to sample at twice the maximum frequency of the signal (also known as Nyquist rate).

A/D conversion

The A/D conversion refers to the conversion of the analog signal to digital form (series of 0s and 1s). This is done by the A/D converter, part of the microcontroller that receives the signal (e.g., from an analog sensor). The resolution of the converter indicates the number of discrete values it can produce over the range of analog values. For instance, the analog sensors we use, operate in the 0-5V range. Resolution in this case refers to the range of integer numbers (usually starting from 0) that represent the latter range. The values are usually stored electronically in binary form, so the resolution is usually expressed in bits. For instance, an ADC with a resolution of 8 bits can encode an analog input to one in 256 different levels, since 28 = 256. The latter suggests that the analog values can be represented between the range of 0 to 255 in case of unsigned integers or from -128 to 127 in case of signed integers, depending on the application and the programming language used.

In the case of Arduino, the internal ADC circuits have 10 bit resolution, meaning that an analog signal from a sensor is transformed to an integer value of range between 0 and 1023.

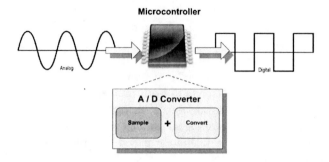

Figure 2-5. An illustration of the digital signal conversion done by a microcontroller

Let's see an example of signal sampling and A/D conversion: Consider an analog sensor that outputs a signal varying from 0V to 5V according to the ambient temperature sensed. When interfaced with the analog port of your Arduino board, the latter range will be converted to integers between 0 and 1023. For instance, a voltage of 3 V will be translated to integer 614. Figure 2-6 illustrates graphically the latter example.

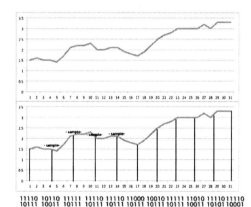

11110 10110 11111 11110 11110 11000 10010 11111 11010 11110 11110
10111 10011 10111 10111 10111 10111 10111 10111 10011 10101 10001

Figure 2-6. An example of sampling and converting an analog signal to digital

Examples of sensors and working principles

This section presents here some of the most used sensors in DIY projects that you will probably come across with, and most likely will use in your own projects too.

Acceleration

One of the most common ways to detect and analyze motion is to use accelerometers. The accelerometer is a sensor that can measure acceleration in one, two or three axes and is often used for measuring velocity, inclination, tilt, vibration and impact in robotics, automobile projects and devices like mobile phones and gaming accessories.

Most accelerometers are Micro-Electro-Mechanical Sensors (MEMS). They are called such because they combine both mechanics and electronics. They contain a small mass etched into the silicon surface of the integrated circuit suspended by small beams. As the sensor moves the acceleration developed moves the mass. The latter movement changes the capacitance of the circuit or generates a piezoelectric effect that can be measured and translated into acceleration.

Figure 2-7 presents an analog accelerometer sensor that measures acceleration in X, Y and Z axis.

Figure 2-7. An analog 3-axis accelerometer sensor. You can see the input Voltage (Vcc), Ground (GND) and sensor outputs for each axis X, Y and Z (image courtesy of Sparkfun).

Capacitive

Accelerometers that implement capacitive sensing output a voltage dependent on the distance between two planar surfaces. One or both of these "plates" are charged with an electrical current. Changing the gap between the plates changes the electrical capacity of the system, which can be measured as a voltage output. This method of sensing is known for its high accuracy and stability. Capacitive accelerometers are also less prone to noise and variation with temperature. They also consume less power, and can have larger bandwidths due to internal feedback circuitry.

Piezoelectric

Piezoelectric sensing of acceleration is natural, as acceleration is directly proportional to force. When certain types of crystal are compressed, charges of opposite polarity accumulate on opposite sides of the crystal. This is known as the piezoelectric effect. In a piezoelectric accelerometer, charge accumulates on the crystal and is translated and amplified into either an output current or voltage.

Piezoelectric accelerometers only respond to AC phenomenon such as vibration or shock. They have a wide dynamic range, but can be expensive depending on their quality.

Piezo-film based accelerometers are best used to measure AC phenomenon such as vibration or shock, rather than DC phenomenon such as the acceleration of gravity. They are inexpensive, and respond to other phenomenon such as temperature, sound, and pressure.

Temperature

When looking for implementations of home automation and Cloud sensor applications you will find that temperature sensors are the most common devices used in such project. The reason is that they are very easy to find, quite cheap, easy ton interface with a microcontroller and measure a basic ambient characteristic, the temperature. They are divided into three different categories: thermocouples, resistance temperature detectors and thermistors (resistive sensors), and temperature-transducing integrated circuits (ICs).

Thermistors are the most commonly devices used in our projects. It varies its resistance as the ambient temperature is changed. Their main measuring range is between -90°C and 130°C. There are two types of thermistors available: Positive temperature coefficient (PTC) and negative temperature coefficient (NTC) thermistors. When temperature increases, the first show an increase of resistance while the latter show a decrease of resistance.

Figure 2-8 presents a thermistor. They usually come with 3 connection pins. One for voltage and ground connection and one for generating the voltage variation readings as the ambient temperature affects the resistance of the circuit.

Figure 2-8. An analog temperature sensor.

Humidity

By humidity we refer to the water vaporized in air (moisture) and is usually measured by relative humidity (RH) which is the ratio of the moisture existing in the air compared to the saturated moisture level at the same temperature and air pressure conditions. On a rainy day you will find the RH will be more than 90%. RH is one of the most common metrics in home automation projects. An example of how to measure the RH in your place is provided in Chapter 8.

RH is measured by special instruments called hygrometers that involve the usage of "wet bulb" and "dry bulb" thermometers. In electronic circuits though we use capacitive humidity sensors that are capable of detecting and quantifying RH by measuring the capacitance change between two non-conductive plates. The change in capacitance is typically 0.2–0.5 pF for a 1% RH change.

Figure 2-9 presents an analog humidity sensor sold by Sparkfun. In this case the sensor provides on the output a voltage variation proportional to the RH measured. In other cases humidity sensors provide directly the RH measurement in % by communicating digitally with the microcontroller.

Figure 2-9. An analog humidity sensor (image courtesy of Sparkfun).

Distance

Measuring distance is quite important in projects dealing with presence (e.g., security systems) and navigation (e.g., robots). Two types of sensors are used in this case; infrared and ultrasound.

Infrared Sensors

An infrared sensor (like the one in Figure 2-10) works by transmitting and receiving back a narrow band infrared beam transmitted from a LED. When the beam is reflected back from an object, the sensor measures the reflection angle and is able to calculate the distance from the latter object. Infrared sensors are very low cost devices but have several drawbacks: there has to be a minimum distance from the transmitting LED to the target so that the receiver can capture the reflection and additionally, the accuracy is highly affected by the distance of the object. The greater the distance the less accurate is the measurement. Therefore, infrared sensors are mostly used to detect objects between 10 and 80 cm away. They are generally not affected by temperature or any other source of noise apart from fluorescence light.

Figure 2-10. An infrared distance sensor. Both the emitting light diode and receiver are visible. Red and black cable are commonly used for voltage and ground respectively. The white cable provides the output voltage that varies based on the object distance.

Ultrasound Sensors

Ultrasound sensors operate based on sound waves propagation: sound waves of higher frequency (above human audible range, thus the 'ultra' name) are transmitted from a source and received back from a receiver (both are parts of the sensor, as in infrared sensors). The total 'time of flight' for the signal is used to determine the distance between the sensor and the targeted object. A great benefit of ultrasound sensors is that they do not require direct contact with the sensed medium-object, however they are very sensitive to temperature, reflections and the angle between the sensor and the target.

To generate ultrasound, piezoelectric crystals are used by applying on them a rapidly oscillating electrical signal. The generated electrical charge causes the crystal to expand and contract with the voltage, thereby generating an acoustic wave. The receiver converts the waves back into voltage using the same method. In addition, depending on the distance measured, measurement is relatively quick (it takes roughly 6ms for sound to travel 1m).

Ultrasound sensors are usually found in automation and automobile projects, like in robots, obstacle avoidance, etc.. Figure 2-11 depicts an ultrasound sensor that can be directly interfaced with your Arduino board. The transmitter and receiver modules are quite visible.

Figure 2-11. An ultrasound distance sensor by Parallax. Very common module used in robotics, can be easily mounted on a circuit or breadboard.

Light

An LDR (light-dependent resistor) also known as photoresistor or photocell is actually a variable resistor that has the ability to modify its resistance according to the intensity of the surrounding light. This means that by using the LDR in a simple circuit that allows you to measure the voltage drop cause by the latter, you can detect and quantify light changes.

LDRs are made of a high resistance semiconducting material. Its operating principle relies on the fact that photons absorbed by the material affect the energy of the electrons that conduct electricity which affects the resistance of the material.

Photoresistors have been initially used for detecting smoke and fire (since both smoke and high temperature can also affect the resistance of the conductive material). They are quite easy to build and low cost and thus are often found in many automation systems, like security alarms, lighting systems, etc. In Figure 2-12 you can see how an LDR sensor looks like.

Figure 2-12. A LDR sensor. It can be connected the same way a resistor does. Its resistance will change depending on the light conditions affecting the current flowing within the circuit.

Orientation

As you know, a compass is a navigational instrument sensitive to the magnetic field of the earth. A typical compass has a magnetic strip, which aligns itself with magnetic north, and from this, orientation can be determined.

An electronic or digital compass is either a magnetometer or a fiber optic gyrocompass, which detects the magnetic directions but without any moving parts. The most common magnetic sensing devices are solid-state "Hall effect" sensors. These sensors produce a voltage proportional to the applied magnetic field and also sense polarity. A digital compass is not free to move the same way that an analogue compass is, as it is typically fixed to a circuit board. Figure 2-13 illustrates a typical digital compass sensor on a breakout board that can be connected directly with your Arduino.

Figure 2-13. An electronic digital compass sensor. Information about orientation is delivered in bits (distinguished by high states) on the output that is connected to the digital input port of the microcontroller. Appropriate code is needed to convert bits to sequence of bytes and then to useful information based on vendor specifications.

Sound

A sensor for detecting sound is, in general, called a microphone. The microphone can be classified into several basic types including dynamic, electrostatic, and piezoelectric according to their conversion system.

In our projects we mostly utilize microphones in order to detect the existence of sound in the environment. We can use sound to trigger switches (e.g., turn on lights with a clap), use ultrasound (for not interfering with background noise) to control actuators like a servo, and in more advanced cases to even perform voice recognition and control outputs of a microcontroller by vocal commands. In Figure 2-14 you can see a complete board with a microphone that can be directly interfaced to your Arduino as an analog sensor.

Figure 2-14. An electric microphone analog sensor with on board circuit for sensitivity control (image courtesy of Seeedstudio).

Electric Current

A current sensor (see Figure 2-15) or electric current transformer (CT) is a device that detects electrical current (AC or DC) in a wire, and generates a signal proportional to it. The generated signal is usually an analog voltage or current but can also be provided through a digital output.

The most common design of CT consists of a length of wire wrapped many times around a silicon steel ring passed over the circuit being measured.

Figure 2-15. An analog clamp-based current sensor. It is clamped around a wire transferring electric current to a device and can be integrated within a circuit the same way a resistor does. The resistance changes according to the current floating in the measuring wire (image courtesy of Seeedstudio).

The most useful application of such sensors is the measurement of power consumption of an electric appliance. To do so, you need some additional circuit that will measure voltage and some code to calculate the overall power consumption in Watts. Instructions and circuit schematics can be obtained by the OpenEnergyMonitor Project (http://openenergymonitor.org).

Actuators

As the name suggests, actuators are devices that act based on a specific condition or event called a trigger. There are many actuators that can be found in electronic circuits but we will mostly focus on those that are most usable in DIY projects like the ones presented in this book. Namely, we will talk about relay switches that allow us to activate circuits remotely (e.g., turn on lights) and servo motors for performing simple motion-involving tasks (e.g., changing the direction of a web cam).

Relay Switch

A Relay switch is basically an electrically operated switch. Current flowing through the coil of the relay creates a magnetic field, which attracts a lever and changes the switch contacts. The coil current can be on or off so relays have two switch positions and most have double changeover switch contacts. Relays are mostly used for allowing one circuit to switch a second circuit that can be completely separate from the first. For instance, a low voltage battery circuit can use a relay to switch a 230V AC mains circuit. There is no electrical connection inside the relay between the two circuits; the link is magnetic and mechanical.

Figure 2-16. A typical relay switch. They usually have 4 connector pins.

The coil of a relay passes a relatively large current, typically 30mA for a 12V relay. However you can use as much as 100mA for smaller relays for your ordinary IoT projects. Most ICs (chips) cannot provide this current and a transistor is usually used to amplify the small IC current to the larger value required for the relay coil.

Relays can have many more sets of switch contacts, for example relays with 4 sets of changeover contacts are readily available. Most relays are designed for PCB mounting but you can solder wires directly to the pins providing you take care to avoid melting the plastic case of the relay.

Relay coils produce brief high voltage 'spikes' when they are switched off and this can destroy transistors and microcontrollers in the circuit. To prevent damage you must connect a protection diode across the relay coil. More details on how to use a relay switch in your projects are provided in Chapter 9.

Servo Motor

A Servo is special motor device that instead of rotating continuously it moves its shaft in specific angular positions according to the input signal. Usually the movement is performed between 0 and 180 degrees. The technique for producing the driving signal is called Pulse Coded Modulation and the duration of the pulse defines the angle of the shaft movement. The servo receives a pulse every 20 milliseconds. A pulse with length of 1.5 millisecond will make the motor turn to the 90-degree position.

A typical servo looks like the one in Figure 2-16. It comes with a 3-pin connector, 2 for providing the appropriate voltage and ground connections and one for the pulse signal that will drive the motor. Arduino comes with the appropriate hardware and libraries that enable it to drive such small servo motors.

Figure 2-17. A typical servo motor for home automation projects. It can control small robotic arms or rotate light objects (image courtesy of Seeedstudio).

Summary

In this chapter we have overviewed the fundamentals of sensor and actuator devices and discussed the most common modules utilized in small automation projects. In chapter 1 we have discussed among others the networking technologies that allow such 'Things' to interact and communicate with the Internet. How about putting everything together into a system that senses, interacts with the environment and communicates with devices and the Internet?

Let's consider the following example as illustrated in Figure 2-18. You could utilize a humidity sensor, a light sensor and a current sensor for measuring the conditions of your room environment and the power consumption of several devices. The sensors can be connected to analog/digital input ports of a microcontroller unit. In addition, you could use a couple of actuators for controlling other devices: a relay switch connected to a lamp and to an air conditioning unit and a servo sensor controlling the window shutters of the room. A WiFi communication module attached to the microcontroller can enable it to communicate directly to the Internet.

You could program the microcontroller (examples of how to program the Arduino, our favorite microcontroller platform are provided in following chapters) to automatically turn on the air conditioning based on humidity conditions and the lights based on how dark or light is in your room. Similarly, the servo motor can control the window shutters to open or close according to the daytime. The current sensor can be used to measure the power consumption by the latter devices.

The interesting part comes when you provide Internet access through the WiFi module. The microcontroller can this way send directly measurements and receive triggers to a web application that you have access to. This allows the following functionality:

- You can access to information about the power consumption and the conditions of your room from everywhere using any Internet-enabled device.

- You can modify thresholds that control when lights will go on-off or when the air conditioner will start or stop.

- You can directly send commands to the microcontroller for turning on-off lights, shutters and the air conditioner on demand.

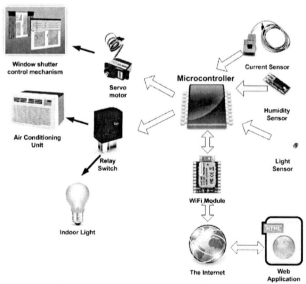

Figure 2-18. Putting Things together: microcontroller, sensors, actuators, wireless modules and the Internet.

You will explore technologies and platforms that provide you with the aforementioned functionality through IoT devices in the following chapters!

3 AN INTRODUCTION TO CLOUD COMPUTING

One fundamental aspect of the Internet of Things is the close connection with Cloud Computing platforms. IoT applications are actually based on Cloud Computing in order to be able to provide storage features with great scalability and great interoperability through open access and direct interfaces for communication and data exchange.

This chapter will give you a basic overview of what the Cloud Computing is, how it is constructed and how users can build applications that communicate with Cloud systems. Finally, the Chapter reviews some of the most popular Cloud-based services that enable users to manage sensor data and build IoT networks.

What is Cloud Computing?

Wikipedia states the following about Cloud Computing as term definition:

"Cloud Computing is the delivery of computing as a service rather than a product, whereby shared resources, software, and information are provided to computers and other devices as a utility (like the electricity grid) over a network (typically the Internet)."

And we agree on that as it best describes the Cloud Computing, the way we become aware of it in everyday Web Services we utilize and the IoT examples we will discuss in this book. Cloud Computing provides various services, like computation, software, data access, and storage, that do not require user knowledge of the physical location and configuration of the system that delivers the services. Figure 3-14 illustrates the concept of Cloud Computing and the communication between services (like applications, data storage and processing resources) and all kinds of computing and communication devices.

Whether you have realized it or not, you're probably already using Cloud Computing. Pretty much everyone with a computer and an Internet connection has been. Gmail and Google Docs are two examples of the many we could name.

Cloud Computing means having every piece of data you need for every aspect of your life at your fingertips and ready for use. Data can be mobile, transferable, and instantly accessible. The key to enabling the portable and interactive you is the ability to synch up your data among your devices, as well as access to shared data. Shared data is the data you access online in any number of places, such as social networks, banks, blogs, newsrooms, paid communities, etc.

Some history

The birth and rise of Cloud Computing is strictly connected to a technology called 'Virtualization'. The latter originates from IBM's research back in the 60's when the company started to develop time-shared, multiuser computers. The main goal back then was to allow many users share the same computational resources on a single computing platform at the same time. With the advances of CPU architectures that allowed many more processes to be executed simultaneously and perform better thread management, virtualization allows users to create and work on several virtual environments on the same machine. The appropriate software was also developed to manage the virtual instances, share the CPU and memory resources appropriately, and take care of issues like system maintenance.

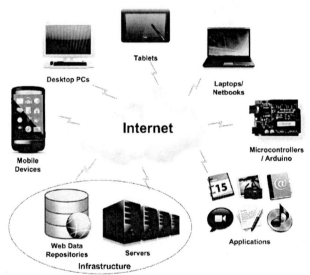

Figure 3-14. An illustration of the Cloud Computing concept. All kinds of computing and communication devices are able to interact with the Cloud and share the same data resources. IoT devices and services are such way a part of the Cloud.

Basic Services and Architectures

A cloud service has three distinct characteristics that differentiate it from traditional application and service hosting. It is provided on demand, it is 'elastic' (meaning that a user can have as much or as little of a service as they want at any given time) and finally, the service is fully managed by the provider. Let's see here what the main components and models of Cloud Computing are.

Cloud Computing Components

A Cloud Computing service (like an online storage service) mainly consists of three basic components: the application, the platform and the infrastructure.

A Cloud application is the main interface with users. It can be unique application (e.g., a storage service) or even represent a group of applications (like Google Docs) run and interacted with via a web browser, hosted desktop or remote client. Users do not need to worry about purchasing and setting up any software (or even the operating system). Instead they pay the service based on their usage.

The Cloud computing platform (mostly the appropriate software components used) is handling any kind of maintenance operations needed (including the activation and integration to the user's application of new hardware resources, like servers or storage elements, for meeting the application and usage requirements. It can be considered as the intermediate layer between the applications hosted on the Cloud and the infrastructure.

The Cloud Infrastructure refers mostly to the hardware layer and includes all the essential computational, storage and networking components that enable the virtualization of the resources and allow the Cloud computing platform to be executed and provide the respective services to users.

Cloud Computing Models and Architectures

There are several types of cloud available, based on the availability of their resources and their usage. Namely, there are public, community-based, hybrid and private clouds.

- Public cloud: Public cloud describes Cloud Computing in the traditional mainstream sense, whereby resources are dynamically provisioned to the general public on a fine-grained, self-service basis over the Internet, via web applications/Web Services, from an off-site third-party provider who bills on a fine-grained utility computing basis.

- Community cloud: Community cloud shares infrastructure between several organizations from a specific community with common concerns (security, compliance, jurisdiction, etc.), whether managed internally or by a third-party and hosted internally or externally. The costs are spread over fewer users than a public cloud (but more than a private cloud), so only some of the benefits of Cloud Computing are realized.

- Hybrid cloud: Hybrid cloud is a composition of two or more clouds (private, community, or public) that remain unique entities but are bound together, offering the benefits of multiple deployment models.
- Private cloud: Private cloud is infrastructure operated solely for a single organization, whether managed internally or by a third-party and hosted internally or externally.

Regarding the architectures and types of services a Cloud provider offers, mainly there are three types of services.

1. Infrastructure as a service: The service providers take care of all the cost of the essential hardware (servers, networking equipment, storage, and back-ups). Users only have to pay to take the computing service. And the users build and manage their own application software. Amazon EC2 is a great example of this type of service.

2. Platform as a service-service: In this case, providers apart from the hardware they also provide a specific platform or a stack of solutions for users (e.g., an application server). It helps users saving investment on hardware and software. Google App Engine and Jelastic provide this type of service. The great benefit of this service is that users need to focus only on the development and deployment of their application and do not worry about setting up operating system and application server issues.

3. Software as a service: Service providers use their own hardware and platform infrastructures in order to provide users with specific software applications, like online storage, email, document applications, etc. Google Docs and Evernote are representative Software-as-a-Service Cloud-based applications.

Benefits of Cloud Computing

So why do you need Cloud Computing and why are you also looking forward to store your sensor data on Cloud-based infrastructures? Some of the most important features and benefits of Cloud Computing are summarized below:

- Removal / reduction of initial costs for hosting a service: With Cloud Computing platforms you do not need to set up yourself the entire appropriate hardware infrastructure (that includes servers, network routers, firewalls, etc.) for hosting your online application and storing your sensor data.

- Reduced administration costs: You do not have to spend time or money on administrating (e.g., setting things up, taking care about backups, etc.) the software and hardware systems for your platform.

- Improved resource utilization: In case you need to build a service and provide it to other users, you do not have to build dedicated systems (web server, database servers) for each separate service. Using Cloud technologies you can reuse much of the existing hardware for hosting several services, even if the latter demand different resources like different operating systems.

- Scalability on demand: How much storage or processing power do you need? What if you start with a couple of sensors monitoring your home and then invite friends to use the same platform and expand to several dozens of sensors? You do not need to extend storage or web server resources yourself; your Cloud service provider does it automatically.

- Quick and easy implementation: You do not need to know anything about setting up a web-based application, setting up and configuring a web server, the database system, making the appropriate connections and configuring several instances to meet scalability requirements. You just focus on the application we need to host on the Cloud and on your data!

- Quality of service and guaranteed availability: The Cloud service provider always guarantees the availability of the hardware and the software. You do not need to worry about hardware failures, power interruptions and network connection on the server side.

- Anywhere Access: Most importantly, the resources of the Cloud (web applications, storage elements) can be accessed from any kind of computational device that has access to the Internet (including your favorite microcontroller platforms).

- Disaster recovery / backup: You do not need to worry about backing up your data or create recovery mechanisms. Everything is taken care of by your provider.

Communicating with the Cloud using Web Services

As described previously, the access to Cloud services has to be easy, direct, open and interoperable, meaning that the provided communication means and programming interfaces (APIs) shall be provided and easy to implement on every platform and developing environment.

Currently, the most open and interoperable way to provide access to remote services and/or enable applications to communicate with each other is to utilize Web Services. The term Web Services is quite self-explanatory: They are services that are provided over the Web, based on particular communication architectures, standards and technologies. They key features of Web services is that they easy to build and deploy and reuse. As a matter of fact, the main reason some one decides to build a web service is for exposing this service to other developers so that the latter can integrate it into their own applications. For example, consider that you have built your own application that manages sensor data and you have deployed it on a Cloud service (check Chapter 7 for more details on how to do that). In order to collect sensor data you need to expose the appropriate functionality to other applications (like your Arduino sketch or your Android mobile app). A Web Service is the perfect mean to do that (you will see on Chapter 7 how easy it is to access the Web Service and send some data to the Cloud).

Web Services can also be seen as interoperable building blocks for developing custom applications.

A Web Service is usually identified by a URI (Unified Recourse Identifier), like a web address.

A Web Service has WSDL (Web Service Description Language) definitions. These are computerized descriptions of what the Web Service can do, where it is located and how it can be used (referred as 'consumed') by the client application. To communicate with Web Services you need to use SOAP messages, which are XML based messages transported over Internet protocols like HTTP, SMTP, and FTP.

Web Services have certain advantages over other technologies:

- Web Services are platform-independent and language-independent, since they use standard XML languages. This means that my client program can be programmed in C++ and running under Windows, while the Web Service is programmed in Java and running under Linux.

- Most Web Services use HTTP for transmitting messages (such as the service request and response). This is a major advantage if you want to build an Internet-scale application, since most of the Internet's proxies and firewalls won't mess with HTTP traffic (unlike CORBA, which usually has trouble with firewalls).

Some Important term you will come across when talking about Web Services:

- Service Processes: This part of the architecture generally involves more than one Web Service. For example, discovery belongs in this part of the architecture, since it allows us to locate one particular service from among a collection of Web Services.

- Service Description: One of the most interesting features of Web Services is that they are self-describing. This means that, once you've located a Web Service, you can ask it to 'describe itself' and tell you what operations it supports and how to invoke it. This is handled by the Web Services Description Language (WSDL).

- Service Invocation: Invoking (or consuming) a Web Service involves passing messages between the client and the server. SOAP (Simple Object Access Protocol) specifies how you should format requests to the server, and how the server should format its responses. In theory, you could use other service invocation languages (such as XML-RPC, or even some ad hoc XML language). However, SOAP is by far the most popular choice for Web Services.

- Transport: Finally, all these messages must be transmitted somehow between the server and the client. The protocol of choice for this part of the architecture is HTTP (Hypertext Transfer Protocol), the same protocol used to access conventional web pages on the Internet. Again, in theory you could be able to use other protocols, but HTTP is currently the most used one.

SOAP and RESTful Web Services

There are currently two options in developing Web Services: the traditional, standards-based approach SOAP and conceptually simpler and newer REST.

SOAP

SOAP stands for Simple Object Application Protocol and has been one of the initial protocol implementations for Web Services provision. It relies on exchanging XML-based messages with a specific format that describe the services (e.g., programming methods) that are provided (this part is called Web Service Description Language and is abbreviated as WDSL) and the data (e.g., sensor readings) communicated between applications.

The basic structure of a SOAP message follows the HTML logic: it includes a header and a body:

```
<env:Envelope xmlns:env="http://www.w3.org/2003/05/soap-
envelope">
    <env:Header>
    <!-- Header information here -->
    </env:Header>
        <env:Body>
        <!-- Body or "Payload" here -->
        </env:Body>
</env:Envelope>
```

The <Header> element is optional, but the <Body> is mandatory.

REST

REST stands for Representational State Transfer. Unlike SOAP, you can get the contents of the service using an HTTP GET request, delete it or update it, using POST, PUT or DELETE actions respectively. It is as if your application that communicates with a REST-based Web services uses whatever the HTTP defines for communicating and exchanging data.

It is a collection of resources, with four defined aspects:

- The base URI for the Web Service, such as http://example.com/resources/
- The Internet media type of the data supported by the Web Service. This is often JSON, XML or CSV but can be any other valid Internet media type.
- The set of operations supported by the Web Service using HTTP methods (e.g., POST, GET, PUT or DELETE).
- The API must be hypertext driven.

A service to get the details of a user called Arduino, for example, would be handled using an HTTP GET to http://example.org/users/arduino. Deleting the user would use an HTTP DELETE, and creating a new one would mostly likely be done with a POST. The need to reference other resources would be handled using hyperlinks (the XML equivalent of HTTP's href, which is XLinks' xlink:href) and separate HTTP request-responses.

RESTful Web Services are also much closer in design and philosophy to the Web itself.

Cloud Computing and IoT

It is clear the Cloud Computing provides great benefits for applications hosted on the web that also have special computational and storage requirements. In addition, it provides easy ways to access them over networks and allows users to focus on the development of the applications. Therefore it is quite sensible that most of the existing IoT platforms relay on Cloud infrastructures. They offer the same user experience with using/building or deploying applications on the Cloud. Users do not have to worry about maintenance of applications, recovery, scalability and networking issues. They just need to configure their sensors to send data over the IoT platforms.

The following section presents the most popular open Cloud-based services for building IoT networks and applications.

Most popular Open Cloud Computing Services for Sensor Management

We will overview here the most popular open Cloud-based applications that offer dedicated services for managing sensor data online. By open we refer to services that allow users to use the basic features (like storing and viewing sensor data online) for free.

The Cosm Service for Internet of Things

Cosm (http://www.cosm.com) has been one of the first on-line database service providers that allow developers to connect sensor data to the Web. It is a real-time data infrastructure platform for the Internet of Things with a scalable infrastructure that enables users to build IoT products and services, and store, share and discover real-time sensor, energy and environment data from objects, devices & buildings around the world.

One of the important features of Cosm that have facilitated its penetration as a IoT cloud service is that the basic usage is free, it is based on an open and easy accessible API and has a very interactive web site for managing sensor data. According to the creators of Cosm, Cosm's website is a little like YouTube, except that, rather than sharing videos, it enables people to monitor and share real time environmental data from sensors that are connected to the internet.

The main capabilities of the service can be summarized into the following:

- Manage real-time sensor & environment data: Since its establishment in 2007 it manages millions of datapoints per day from thousands of users, organizations & companies around the world. Cosm can store, convert and distribute data in multiple formats (JSON, XML and CSV), which makes it interoperable with the most established web protocols application development standards. All feeds include contextual metadata (timestamps, geolocation, units, tags) that add value to datastreams.

- Graph, monitor & control remote environments: Cosm allows users to embed real-time graphs and widgets in their website, analyze and process historical data pulled from any public Cosm feed. Additionally, to send real-time alerts from any datastream to control your scripts, devices & environments. There are also available configurable tools that include a zoomable graph, a mapping/tracking widget, an augmented reality viewer, SMS alerts & apps for various smartphones.

- With a rapid development cycle & dozens of code examples & libraries, Cosm's 'physical-to-virtual' API makes it possible to build applications that add value to networked objects & environments. Cosm handles the scalability & high-availability required for complex data management.

Figure 3-15 shows a screenshot from Cosm's main web interface for viewing 'feed' information and sensor data history graphs. The 'feed' refers to the environment you are monitoring through your sensors. For instance, feed can refer to your room where a temperature sensor is placed.

Cosm supports a great number of software libraries for developing custom applications that communicate with it. Currently, Java (through Processing), Ruby, Python, .Net, Perl, PHP, LabView, C/C++, Visual Basic, JavaScript are supported through available APIs. Since Cosm is based on RESTful Web Services, it is clear that it can be embedded in any application development environment that can support the use of the latter Web Services.

A simple example of how to invoke the Cosm Web Service is the following URL that requires the status of a specific feed (28602) about three sensors (datastreams):

```
https://api.cosm.com/v2/feeds/28602.csv?key=9eIViKnjsmiP6ST4
T4aTzlmdw8x5ASqkbU_Pd86qtAg
```

The response is a CSV formatted message containing the current sensor values as the following:

```
0,2011-11-29T07:53:43.584864Z,18.70
1,2011-11-29T07:53:43.584864Z,43.59
2,2011-11-29T07:53:43.584864Z,597
```

You can check the latter by calling the URL within your browser (or use this shortcut instead: http://tinyurl.com/dyeuj4v). You get back a CSV response since this is the format defined in the HTTP request. Try yourself modifying the URL to receive the JSON formatted response and see the difference.

Already a great number of microcontroller and embedded devices support the Cosm service though appropriate software libraries that allow easy communication with the service. Some of them are listed here:

- Current Cost Bridge:
- Arexx: Wireless sensing system with remote sensors and LAN receiver

These two are Cosm-ready consumer products, meaning that they are commercial products that natively support storing sensor data on Cosm without the need from user to program them appropriately.

- The Nanode: Nanode is an open source Arduino-like board that has in-built web connectivity. It connects to a range of wireless, wired and Ethernet interfaces. It allows you to develop web based sensor and control systems - giving you web access to six analogue sensor lines and six digital I/O lines.
- OpenGear ACM 5000 Console Server: It is a Low-power embedded Linux device server with USB, serial, data ports and Ethernet –(comes also with optional WiFi and 3G connectivity). It includes temperature, dust and water sensors. Includes Custom Development Kit for rapid application bring up and kernel customization.
- Arduino: Our favorite microcontroller platform, lots of examples and further introduction on the following chapters of the book!
- mBed microcontroller: The mbed Microcontroller is made for prototyping, and comes in a 40-pin 0.1" pitch DIP form-factor so it's ideal for experimenting on breadboard, stripboard and PCBs. It supports lots of interfaces, so you can connect it to all sorts of input and output circuits and modules.

Figure 3-15. The Cosm web application that allows users to set feeds and data points, visualize them and setup triggers.

More details about the Cosm service and how to use it with Arduino are presented in Chapter 8.

The Nimbits Data logging Cloud Server

Nimbits (http://www.nimbits.com) is a data processing service you can use to record and share sensor data on the cloud. It is a free, social and open source platform for the Internet of Things.

With Nimbits, you can create data points on the cloud and feed your changing numeric, text based, GPS, JSON or xml values into them. Data points can be configured to perform calculations, generate alerts, relay data to social networks and can be connected to SVG process control diagrams, spreadsheets, web sites and more. Explore this site to learn how to use the open source software development kit, Chart API, Portal Server and more. Figure 3-16 shows a screenshot from the main Nimbits web interface for managing sensor data.

There are many ways to record historical values into Nimbits. You can either write software interfaces using the SDK, or record data manually to points using the Portal, Data Acquisition Studio, or the Android Data Logger.

Nimbits offers the following sensor data processing features:

- Data point Compression: Compression in Nimbits allows you to filter out noise in oncoming data. This is very useful if you are recording a reading that doesn't change much. For example, a temperature data point that had a previous value of 70, and a compression setting of 1 will ignore any newly recorded value that that is between 71 and 69. Compression is useful for filtering out noise, such as a reading that goes up and down by a small value (i.e. .01) and you only want to record significant changes or if the incoming value is usually the same and it is inefficient to continuously record the same value over and over.

- Alert states of data points: A point can be in one of four states: Normal All is well. High: The last recorded number value was higher or equal to the high alert setting. Low: The last recorded number value was lower or equal to the high alert setting. Idle: The point has not received a new value in a set number of time, based on the idle setting. When a point goes into an Alert State, it will attempt to notify you in ways set in the point's properties, through Email, Instant Message, Facebook, Twitter and when using the portal or android app, point icons change color based on the alert state.

- Calculations: When a DataPoint is written to with a new value, the event of recording a value can trigger a calculation. The formulas basic arithmetic will be executed and the resulting value will be recorded into the target point – all compression, alerts and other functionality will then be applied to the new value as if it was a newly recorded value. A DataPoint can be a trigger, parameter or

53

target for a formula. If a point is a trigger and a new value is written to it the formula assigned to that data point will be executed. After the formula executes successfully, the resulting double value is stored in the target point and all Record Value settings are processed (Such as Alerts, Compression etc.)

- Data Expiration: Users might need to configure a data point to automatically delete data that is older than a certain date. This can be good housekeeping to clean out old, scale code. It can also provide good documentation on how long you keep your data stored. Depending on your industry, you may be required to keep data up to a certain date. Based on your point settings, Data can be permanently deleted when it reaches a certain age. Also, incoming data with a timestamp older than the expiration setting will be ignored by the system.

Figure 3-16. The Nimbits main web application environment. Users can define data points for storing sensor data, view graphs of sensor history and use the service features, like alarm settings and data calculation.

Nimbits can obviously support any device that can be programmed to make a HTTP POST request with a JSON message and currently provides implementation examples for the Arduino and the Parallax Stamp microcontrollers. We will check the Nimbits platform in details in Chapter 9.

ThingSpeak Internet of Things
ThingSpeak (https://www.thingspeak.com) is another open source "Internet of Things" application and API to store and retrieve data from things using HTTP over the Internet or via a Local Area Network. With ThingSpeak, users can create sensor logging applications, location tracking applications, and a social network of things with status updates. In addition

to storing and retrieving numeric and alphanumeric data, the ThingSpeak API allows for numeric data processing such as time scaling, averaging, median, summing, and rounding. Each ThingSpeak Channel supports data entries of up to 8 data fields, latitude, longitude, elevation, and status. The channel feeds support JSON, XML, and CSV formats for integration into applications. The ThingSpeak application also features time zone management, read/write API key management and JavaScript-based charts (see Figure 3-17).

To read and write to a ThingSpeak Channel, your application must make requests to the ThingSpeak API using HTTP requests. Each ThingSpeak Channel allows for 8 fields of data (both numeric and alphanumeric formats), location information, and a status update. Each entry is stored with a date and time stamp and is assigned a unique Entry ID (entry_id). After the data is stored, you can retrieve the data by time selection or by Entry ID. In addition to storing and retrieving numeric and alphanumeric data, the ThingSpeak API allows for numeric data processing such as time scaling, averaging, median, summing, and rounding. The channel feeds supports JSON, XML, and CSV formats for integration into applications.

Here is an example HTTP POST to ThingSpeak:

```
POST /update HTTP/1.1
Host: api.thingspeak.com
Connection: close
X-THINGSPEAKAPIKEY: (Write API Key)
Content-Type: application/x-www-form-urlencoded
Content-Length: (number of characters in message)
```

ThingSpeak includes applications, called Apps, that make it easier for devices to access resources on the web such as social networks, Web Services, and APIs:

- ThingTweet: ThingTweet is a Twitter Proxy that allows a device to send status updates to Twitter via simple API calls to the ThingTweet App.
- ThingHTTP: ThingHTTP is for connecting things to Web Services via HTTP requests. ThingHTTP supports GET, POST, PUT, and DELETE methods, HTTP/1.0, HTTP/1.1, SSL, custom HTTP headers, and Basic Authentication.
- TweetControl: TweetControl allows you to monitor your Twitter stream for a specific hashtag and associate it with a control action. The TweetControl App uses the Twitter Stream API and operates in real-time. You can create social controls via Twitter without having to run a dedicated server listening to the Twitter firehouse.

Figure 3-17. The channel visualization web application of ThingSpeak. Users can select channels (representing sensor data streams), view generated data history graphs and also receive html code for embedding graphs in their own web applications.

The iDigi Device Cloud

iDigi Platform is a machine-to-machine (M2M) platform-as-a-service. iDigi Platform lowers the barriers to building secure, scalable, cost-effective solutions that seamlessly tie together enterprise applications and device assets. iDigi Platform manages the communication between enterprise applications and remote device assets, regardless of location or network. It makes connecting remote assets easy, providing all of the tools to connect, manage, store and move information across the near and far reaches of the enterprise.

The platform includes the device connector software (called iDigi Dia) that simplifies remote device connectivity and integration. It allows the management (configure, upgrade, monitor, alarm, analyze) of products including ZigBee nodes. The application messaging engine enables broadcast and receipt notification for application-to-device interaction and confirmation. There are also cache and permanent storage options available for generation-based storage and on-demand access to historical device samples.

Digi Dia (Device Integration Application) is software that simplifies connecting devices (sensors, PLCs, etc.) to communication gateways. Dia includes a comprehensive library of plug-ins that work out-of-the-box with common device types and can also be extended to include new devices. Dia's unique architecture allows the user to add most of these new devices in under a day.

iDigi Dia is designed upon a tested architecture that provides the core functions of remote device data acquisition, control, and presentation between devices and information platforms. It collects data from any device that can communicate with a Digi gateway, and is supported over any gateway physical interface. Digi Dia presents this data to upstream applications in fully customizable formats, significantly reducing a customer's time-to-market. Written in the Python programming language for use on Digi devices, iDigi Dia may also be executed on a PC for prototyping purposes when a suitable Python interpreter is installed.

The SensorCloud

MicroStrain's SensorCloud is a unique sensor data storage, visualization and remote management platform that uses Cloud Computing technologies to provide data scalability, rapid visualization, and user programmable analysis. Originally designed to support long-term deployments of MicroStrain wireless sensors, SensorCloud now supports any web-connected third party device, sensor, or sensor network through a simple OpenData API, like the Arduino.

The core SensorCloud features include:

- Virtually unlimited data storage with redundant reliability, ideal for collecting and preserving long-term sensor data streams
- Time series visualization and graphing tool with exceptionally fast response, allows viewers to navigate through massive amounts of data, and quickly zero in on points of interest
- MathEngine feature allows users to quickly develop and deploy data processing and analysis apps that live alongside their data in the cloud
- Flexible SMS and email alert scripting features helps users to create meaningful and actionable alerts

SensorCloud is useful for a variety of applications, particularly where data from large sensor networks needs to be collected, viewed, and monitored remotely. Structural health monitoring and condition based monitoring of high value assets are applications where commonly available data tools often come up short in terms of accessibility, data scalability, programmability, or performance.

The main features of the web interface that allows users to manage data include the following:

- Visualize: SensorCloud is unique in that it leverages several new Cloud Computing technologies to make it easier to work with extremely large data sets. Its graphing functions utilize parallel computing and client side caching to accelerate responses to user graphing requests. The result is a high-performance web data visualization tool that typically generates plots in under a second, and allows users to quickly navigate through gigabyte, terabyte, and even petabyte sized data sets.

- Analyze: SensorCloud is also unique in that it is the first web-based sensor data aggregation platform that provides a flexible online analytics tool supporting user-developed apps. SensorCloud's MathEngine allows users to quickly develop and deploy data processing apps in either Python and Octave (an open-source, code compatible MATLAB® alternative), and will soon support the popular R statistical language, giving users programming language flexibility to best suit their needs. Users can either upload their code or use an online editor to develop a wide range of data processing apps, from simple one-time scripts for filtering and statistical analysis, to advanced, continuously-running online algorithms for health monitoring and prognostics.

- Collaborate: Cloud Computing continues to change how we produce, consume, and share information. In similar respect, SensorCloud changes how teams share and work with sensor data. With secure HTTPS/SSL web access standard, SensorCloud simplifies data sharing and analysis for team members spanning multiple locations, it helps groups better coordinate event responses with flexible alerts, and it allows teams to collaborate on code development for MathEngine apps. Data owners can also expand their audience by sending invites to domain experts to view their data set, assist with analysis, and develop advanced, custom-tailored data processing applications.

- Scale: Cloud Computing has reduced barriers and costs associated with access to powerful computing capabilities previously reserved for expensive enterprise systems, while at the same time bringing new tools and techniques that exceed enterprise system performance and scalability. MicroStrain's SensorCloud seeks to transition these new and unique cloud capabilities into tools that make it easier to collect and visualize large quantities of sensor data, and extract meaningful insight with user-programmable online analytics.

The OpenData API allows users to upload sensor data from any web-connected source or platform, and download selected or entire data sets. It includes features like Secure data upload and download using HTTPS & SSL a fully REST compliant API, example code for common languages and platforms (python, Java, C#, C++, Labview, iPhone, Android). Data download currently supports CSV and XDR file formats.

The Alerts feature allows users to create custom email and SMS text message alerts for monitoring data when they exceed values and events of interest.

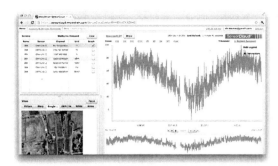

Figure 3-18. The graph visualization web application of SensorCloud. Users can browse devices and channels and view a zoomable graph of sensor data history.

The MathEngine allows users to process vast quantities of sensor data in the cloud, and on the fly. Users can either create simple custom scripts for basic operations such as averaging and filtering, or, they can implement advanced data processing algorithms for generating condition indicator metrics such as metal fatigue or bearing failure. More than just a data post-processing tool, MathEngine output can be fed directly to the Alerts App, enabling fully customized & powerful alert scripting.

SensorCloud organizes data into a hierarchy of different components. The Device is at the top-level component. A device contains Sensors. Each Sensor may have one or more channels and each channel may have one or more datastreams. A datastream is where the sensor data is actually stored.

A List of available platforms

The number of available platforms that feature data logging on the Cloud for sensor devices is growing continuously in a rate that looks hard to track! There are even yet so many that it would take a new book to present them. The following list overviews the ones presented here and some more that are worth exploring.

Table 3-1. Overview of the Cloud IoT Platforms.

Name	Features	Free | Open Source	URL
Cosm	Visualize and store sensor data online	Yes | No	http://www.cosm.com
Nimbits	Data Logging on the Cloud	Yes | Yes	http://www.nimbits.com

ThingSpeak	Visualize and store sensor data online	Yes	Yes	https://www.thingspeak.com
iDi	Device Cloud platform	Yes	No	http://www.idigi.com
SensorCloud	Visualize and store sensor data online	Yes	No	http://www.sensorcloud.com
Open.Sen.Se	Internet of Everything platform	Yes	No	http://open.sen.se
Exosite	Platform and Portals for Cloud-based data and device management	Yes	No	http://www.exosite.com
EVRYTHNG	Software Engine and Cloud Platform	Yes	No	http://evrythng.net
Paraimpu	Social tools for thins	Yes	No	http://paraimpu.crs4.it
Manybots	Collect and manage information from various devices	Yes	No	https://www.manybots.com
Lelylan	Focused on home automation and monitoring	Yes	--	http://lelylan.com

Summary

We have discussed so far about the Internet of Things and what kind of sensors you can connect to the Internet. This chapter has introduced the basic concepts of Cloud Computing, the most fundamental component of IoT networks and applications. You have been presented with the basic services and modules of Cloud Computing and the major benefits of this new concept in managing and storing online information. The best way to communicate your own applications with Cloud infrastructures is through Web Services and specifically RESTful Web Services that are so easy and lightweight to use that are the most suitable for microcontrollers and devices. We have also discussed about the most popular Cloud-based IoT services for managing sensor data online.

The following chapter introduces the most popular open-hardware microcontroller platform, the Arduino. You will learn how to program it and use some of the most advanced libraries that exist for building better and most effective embedded applications.

PART II

THE ARDUINO PLATFORM

"The real key is remote management. It integrates into the platform. A Microcontroller allows an administrator to wake the system up, even when it's lights out."

Martin Reynolds

4 THE ARDUINO MICROCONTOLLER PLATFORM

We have discussed so far about the basic concepts of the Internet of Things and how your devices can be connected to it, how sensors work and what they can sense from your environment and what the Cloud computing is and how it can assist on better managing your sensor data.

This chapter will present and teach you how to use the Arduino, our favorite open source Microcontroller platform. You will see what are the available boards you can use, how to set up and use the development environment (also called IDE), and how to write your own programs (called 'Sketches') that will give life to your board and make it do something useful and interesting! Finally, we will discuss on ways you can extend your Arduino board by using one or more of the numerous extension shields that exist, how use the Microcontroller out of the board and most importantly, how to use external libraries that will help you program it for the IoT.

Let's first have a look about Microcontrollers.

Microcontrollers

A Microcontroller (also abbreviated as MCU – from Microcontroller Unit) can be considered as a small computer on a single integrated circuit (you might understand it better as a 'microchip') containing a processor core, memory, and programmable input and output peripherals. As we discussed in previous chapter, typical input and output devices include switches, relays, solenoids, LEDs, small or custom LCD displays, radio frequency devices, and sensors for data such as temperature, humidity, light, etc. The Program memory (in the form of Flash memory) is also often included on chip, as well as a typically small amount of RAM. The majority of Microcontrollers in use today are embedded in any kind of machines and devices, such as automobiles, telephones, appliances, and peripherals for computer systems. These are called embedded systems. Figure 4-20 illustrates a Microcontroller, which is also the electronic brain of this chapter's topic, the Arduino. Microcontrollers are mostly designed for and used in such embedded applications in contrast to the Microprocessors that are used in PCs.

The Microcontroller is a very common component in modern electronic systems. Its use is so widespread that it is almost impossible to work in electronics without coming across it. The main components of a Microcontroller unit are as illustrated in Figure 4-19:

- I/O (Input/Output) ports. These are usually pins that collect and generate digital signals to other circuits. They are the interfaces of the Microcontroller with the outer world. Sensors and actuators as well as other devices that communicate with the Microcontroller are connected to these ports. Since these are digital ports (this is also why only digital signals can be communicated) the Microcontroller receives information in series of Bits and Bytes.
- The CPU: The main processing unit where all calculation are made based on the programming and the interaction with the external circuits.
- The Memory: It includes the program that is being executed and is also available for storing some information that the program generates (e.g., readings from input ports). It is usually very limited and can cause issues in large programs and in cases memory is not well managed.
- A serial line from the microprocessor (Transmit or TX) and a serial line to the microprocessor (Receive or RX) allowing serial data in the form of a bit stream to be transmitted or received via a two wire interface.

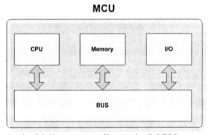

Figure 4-19. A typical Microcontroller unit (MCU) structure. The basic components are the CPU, the Memory and the I/O ports.

Figure 4-20. A Microcontroller chip. The communication with the rest of the board/circuit components as well as the powering is done through its pins. This is the ATMega328 40-pin chip.

Most Microcontrollers will also combine other devices such as:

- A Timer module to allow the Microcontroller to perform tasks for certain time periods.
- An ADC (Analog to Digital Circuit) or converter that allows the Microcontroller to accept analogue input data for processing.

The most popular types of Microcontrollers are the Parallax Propeller, Basic Stamp, PICAXE, ARM processors and Atmel AVRs. Arduino is using the Atmel ATMega series and the most common one is the ATMega328 chip (see Figure 4-20).

Programming Microcontrollers

The main function of a Microcontroller is what the name actually suggests: to control things through the I/O interfaces. In order to instruct the Microcontroller what to control and how, you need to program it. Microcontrollers were originally programmed only in assembly language, but various high-level programming languages are now also in common use to target Microcontrollers. Such languages are either designed specially for programming Microcontrollers, or versions and variations of general-purpose languages such as the C programming language are commonly used. Microcontroller vendors often make tools freely available to make it easier to use their hardware.

In most cases you need particular hardware in order to program the Microcontroller. This hardware is also known as Microcontroller programmer. The main issues that one might face with Microcontroller programming are two: a) the need for special hardware that often does not come at a reasonable price (at least for hobbyists and prototype usage) and b) the difficulty in programming which can be translated into long programming times or too many interconnections between the controller and the programmer itself.

Many of these problems are addressed by a 'Bootloader'. The Bootloader is a small program that has been loaded on to the Microcontroller on your board. This piece of software is programmed in the program memory of the Microcontroller just once, using a conventional programmer. After this, the Microcontroller can be programmed without a programmer. Once in the Microcontroller, the Bootloader is such programmed that each time after reset it starts running like any conventional program. What it does however is different from a regular program. First of all, depending on what type of Bootloader it is, it starts "listening" for incoming bytes via a specific interface. For instance, a UART Bootloader will listen to the UART buffer of the micro, checking for incoming bytes. If the bytes start arriving, the Bootloader will grab them and write them in the program memory in the sequence it receives them and at predefined locations. Once all bytes have been received, the Bootloader executes a jump at the start of the memory zone it has received and then the "normal" program starts running.

The Arduino platform offers mainly three things that have removed completely any complexity and made programming Microcontrollers easy and accessible even to inexperienced users: a) An open-source Bootloader, b) open-schematic boards that can be used out-of-the box for programming and building custom projects, c) A development environment for writing programs and uploading them on the boards. In addition, there is a great expanding community behind Arduino that supports it and evolves it. Let's see some more details about the Arduino platform.

The Arduino Platform

According to its creators, Arduino is an open-source electronics prototyping platform based on flexible and easy to use hardware and software. It can be used to develop custom electronic projects, taking inputs from a variety of switches and/or sensors, and controlling a variety of outputs (lights, motors, etc.) In simple terms, Arduino can be the brain of your every single project. Arduino programs can be stand-alone (can be executed only on the board), or they can be used to forward or receive information from applications running on your computer (using for example Flash or Java/Processing). The boards can be assembled by hand or purchased preassembled. The software used to program the board is open-source and can be downloaded for free.

The main features that are also the strong benefits of Arduino are the following:

- It is low cost: Arduino boards are relatively inexpensive compared to other Microcontroller platforms. Almost all Arduino modules can be assembled by hand, and even the pre-assembled ones cost less than $50
- It is Cross-platform: The Arduino software runs on Windows, Macintosh OSX, and Linux operating systems.
- Simple, clear programming environment - The Arduino programming environment is easy-to-use for beginners, yet flexible enough for advanced users to take advantage of as well. For teachers, it's conveniently based on the Processing programming environment, so students learning to program in that environment will be familiar with the look and feel of Arduino
- Open source and extensible software- The Arduino software and is published as open source tools, available for extension by experienced programmers. The language can be expanded through

C++ libraries, and people wanting to understand the technical details can make the leap from Arduino to the AVR C programming language on which it's based. Similarly, you can add AVR-C code directly into your Arduino programs if you want to.

- Open source and extensible hardware - The Arduino is based on Atmel's ATMEGA8 and ATMEGA168 Microcontrollers. The plans for the modules are published under a Creative Commons license, so experienced circuit designers can make their own version of the module, extending it and improving it. Even relatively inexperienced users can build the breadboard version of the module in order to understand how it works and save money.

Arduino is not a Microcontroller itself. It is a platform for open source prototyping that consists of hardware boards (that include Microcontrollers) and a programming environment.

Another great advantage of Arduino is that if you can test, build and deploy embedded projects very fast. You can use an Arduino board to build a prototype, test it and then simply remove the chip out of the board and use it into your own circuit board creating your own embedded device. You can also get single ATmega chips and program them using an Arduino board. Note that chips must be pre-programmed with the Arduino Bootloader (software programmed onto the chip to enable it to be used with the Arduino IDE). It is also possible to program a chip using a second Arduino.

The Boards

Arduino exists in several different board variants. The official Arduino team designs most of them. You can also find various clones of the most common boards (like the Uno or the Mega) by other vendors.

An Arduino board consists of an 8-bit Atmel AVR Microcontroller with all the essential components that are necessary for allowing you to program it and interact with sensors and other circuits. An important aspect of the Arduino is the standard way that connectors are exposed, allowing the CPU board to be connected to a variety of interchangeable add-on modules (known as shields). Official Arduinos have used the megaAVR series of chips, specifically the ATmega8, ATmega168, ATmega328, ATmega1280, and ATmega2560. Most boards include a 5 Volt linear regulator and a 16 MHz crystal oscillator. An Arduino's Microcontroller is also pre-programmed with a boot loader that simplifies uploading of programs to the on-chip flash memory, compared with other devices that typically need an external chip programmer.

The various board versions differ in size and capabilities in terms of processor and number of I/O pins that can be utilized. Figure 4-21 presents the most popular official Arduino boards.

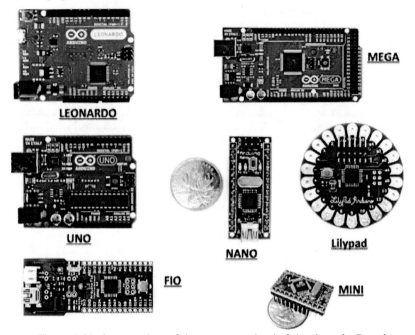

Figure 4-21. An overview of the most popular Arduino boards. Boards differentiate in size, features and capabilities (board images courtesy of the Arduino community).

A brief overview of each board is listed here:

- Leonardo: This recently introduced board is based on the ATmega32u4. Comes with 20 digital input/output pins. 7 of them can be used as PWM outputs and 12 as analog inputs. The board contains everything needed to support the microcontroller (16 MHz crystal oscillator, a micro USB connection, a power jack, an ICSP header, a reset button, a LED, etc.). The Arduino Leonardo has built-in USB communication, (as opposed to other boards). This allows the Leonardo to appear to a connected computer as a mouse and keyboard, in addition to a virtual (CDC) serial / COM port.

- Uno: Based on the ATmega328 chip, it has 14 digital input/output pins, 6 analog inputs, a 16 MHz crystal oscillator, a USB connection, a power jack, an ICSP header, and a reset button. It is comes as a ready-to-use board, meaning that it contains everything needed to program it and use it. The Arduino Uno can be powered via the USB connection or with an external power supply (you don't need to worry how the power source is selected, it is done automatically).

- Mega: It is based on the ATmega2560 and features 54 digital input/output pins, 16 analog inputs, 4 UARTs (hardware serial ports), a 16 MHz crystal oscillator, a USB connection, a power jack, an ICSP header, and a reset button. It is considered as a more advanced version of Uno, used mainly for the greater number of I/O ports. Similar to Uno it can be used directly out-of-the box.

- Nano: This one is a small but complete board based on the ATmega328 (Arduino Nano 3.0) orATmega168 (Arduino Nano 2.x). It has more or less the same functionality of the Arduino Uno, but in a different package. It lacks only a DC power jack, and works with a Mini-B USB cable instead of a standard one. It is also considered breadboard-friendly because of its size and pin layout that allow it to sit on a breadboard.

- Lilypad: This board is designed for wearable projects because it can be sewn to fabric and similarly mounted power supplies, sensors and actuators with conductive thread. The board is based on the ATmega168V (the low-power version of the ATmega168) or the ATmega328V. It provides the same I/O pins with Uno, but needs additional hardware for programming (FTDI USB adapter).

- Mini: It is the smallest version of an Arduino board one could get. Still it has all the functions of Uno (regarding memory and I/O pins) since it can come with the same Microcontroller chip. Like Lilypad you need an external adapter to program it and can be placed on a breadboard like the Nano.

- Fio: This is one of the best options for IoT projects that require connectivity. It is based on the ATmega328P Microcontroller, operates at 3.3V and 8 MHz. It has 14 digital I/O pins, 8 analog inputs, an on-board resonator, a reset button, and holes for mounting pin headers. It has connections for a Lithium Polymer (LiPo) battery and includes its own LiPo charge circuit over USB. It is suitable for wireless applications since it contains an XBee socket.

The XBee socket is a specially designed socket that accepts 'Bee' series connection modules (Figure 4-22). The latter are communication modules designed as pluggable extensions to Arduino boards. They have a specific size and layout and are powered directly from the Arduino board through the socket. The socket also provides direct communication with the Rx/Tx pins of the Arduino. The XBee was initially introduced by Digi as a wireless communication module for ZigBee support. Once you have a Bee shield (or a Fio) you can plugin a Bee module and start working without worrying about connection and powering. Currently, there are available Bee modules that provide ZigBee, Bluetooth and WiFi connectivity. One great aspect of the Arduino Fio and the XBee modules is that by using a modified USB-to-XBee adaptor (such as XBee Explorer USB), you can upload sketches wirelessly and program your Arduino remotely!

Figure 4-22. A Bluetooth Bee module on a XBee shield with two Bee sockets available

Other Arduino boards are the Arduino Mega ADK (similar to Mega but with an additional USB host interface to connect with Android based phones), the Arduino BT (similar to Uno but with a built-in Bluetooth module) and the Arduino Ethernet Board (similar to Uno but with a built-in Ethernet port. Requires also external USB module for programming like the Nano).

It is clear that the Arduino team and the rest Arduino vendors are moving towards providing network capabilities to the boards facilitating the connection to the Internet and the communication with Cloud systems!

The Anatomy of an Arduino Board

As mentioned previously the Arduino board consists of an AVR Microcontroller (the brain of the board), usually an ATmega328, appropriate electronics for it function and additional components for letting you to program it and also communicate with sensors and other circuits like communication modules.

Figure 4-23 presents an Arduino Uno board (facing upwards) among with an annotation of the major board components. You can see a 40-pin version of the Microcontroller chip mounted on a chip socket near the center of the board, the power jack, a USB port, a reset button, a serial programmer and two rows of I/O pins on the upper and lower part of the board. The reset button resets the board and re-starts the sketch execution from the beginning. The serial programmer can be used in case you wish to program the board using an external programmer (which will also allow you to modify/update the Bootloader).

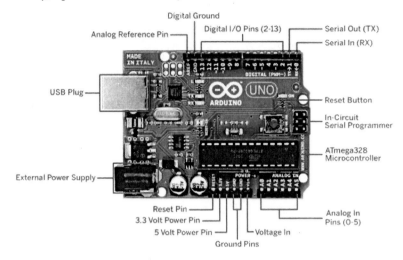

Figure 4-23. The anatomy of the Arduino Uno board. You can see among others, the main pins for attaching analog and digital sensors, the power and ground pins, the power supply and the ATmega328 controller on the board (image courtesy of Arduino).

You can power the Arduino via the USB connection or by using an external power supply. The power source is selected automatically. In case of external power source you need to connect an adapter by plugging a 2.1mm center-positive plug into the board's power jack. Alternatively, leads from a battery can be inserted in the Gnd and Vin pin headers of the POWER connector. The board can operate on an external supply of 6 to 20 volts. If supplied with less than 7V, however, the 5V pin may supply less than five volts and the board may be unstable, so the recommended power range is between 7 and 12 volts.

On the lower – right part of the board you will notice the analog input pins (0-5). These are the pins that you can use to connect analog sensors and read their inputs directly from your Arduino sketch. On the left side there is another series of 6 pins. The reset pin is a digital input port. You can control for instance an actuator there and when set to HIGH state (i.e. when 5V voltage goes through) the pin will be activated resetting the board. The following two pins generate 3.3V and 5V respectively and can be used so that you can power the sensors and/or other components of your board.

Next to them there are two Ground pins and a voltage in pin. As mentioned before, you can use the pair of one Ground pin and the Voltage in pin to power your board through a battery.

The upper part of the board contains the digital I/O pins and the serial communication pins. The latter are the pair annotated as Rx and Tx on the board. You will use these pins in order to communicate with modules that use the serial protocol like Bluetooth modems, LCDs, etc. Most shields also that provide communication abilities to your Arduino are utilizing the latter pins. Moving to the left there are 12 digital I/O pins. These can be used to read digital (sequence of HIGH/LOW states) data from modules like switches or digital sensors or to control actuators like a relay switch. Their operating state (input or output) is defined within the sketch code. There is also another ground pin (quite useful if you want to directly plug something like a LED to a digital pin) and finally the analog reference pin. The latter is used in cases you need to have a reference voltage in your circuit, i.e. a voltage you are sure what the value is and it is not by default 3.3V or 5V. The voltage of the analog reference pin is set programmatically from your sketch code. The available options are:

- DEFAULT: the default analog reference of 5V or 3.3V depending on the board type (e.g., for UNO it is 5V)
- INTERNAL: a built-in reference, equal to 1.1V on the ATmega168 or ATmega328 and 2.56 volts on the ATmega8(not available on the Arduino Mega)
- INTERNAL1V1: a built-in 1.1V reference (Arduino Mega only)
- INTERNAL2V56: a built-in 2.56V reference (Arduino Mega only)
- EXTERNAL: a custom voltage applied to the AREF pin (between 0 and 5V only)

The Development Environment

The Arduino IDE is a graphical cross-platform application written in Java, derived from the IDE for the Processing programming language and the Wiring project. It is characterized by great simplicity (compared to what you might have seen in other programming IDEs) so that people even totally unfamiliar with software development can use it. It includes a code editor with features such as syntax highlighting, brace matching, and automatic indentation, and is also capable of compiling and uploading programs to the board with a single click (see Figure 4-24). There is typically no need to edit makefiles or run programs on the command line. Since it is Java-based it can be installed and executed in every computing environment, Windows, MacOSX and Linux.

The Arduino IDE offers mainly the following features to the users:

- Manage programs (called sketches): Edit new programs in the code editor, load and save them.
- Compile the sketches and view any success or error messages on the message console.
- Upload sketches to the Arduino board user has connected to the computer.
- The Serial Monitor: A graphical interface that displays all the information coming from the serial communication with the Arduino board and that also allows users to send information back to the Arduino.

The IDE comes also with several examples preinstalled so that the user can browse them and upload them directly to an Arduino board. The examples demonstrate the basic functionalities of the Arduino.

Setting up the IDE

For the latest IDE go to the Arduino download page at http://arduino.cc/en/Main/Software and obtain appropriate the version for your operating system.

Installing on Windows XP
Once you have downloaded the latest IDE, unzip the file and open it. You will see the Arduino files and sub-folders inside. Next, plug in your Arduino using the USB cable and ensure that the green power LED (labeled PWR) turns on. According to the board you are using you will get a relative windows notification like "Found new hardware: Arduino Uno" and the Found New Hardware Wizard will appear. Click next and Windows will attempt to load the drivers. This process will fail since there are no default USB drivers for the board available in the Windows system. Right-click on the My Computer icon on your desktop and choose Manage. The Computer Management window will open up. Now go down to Event Manager in the System Tools list and click it. In the right hand window, you'll see a list of your devices. The Arduino Uno will appear on the list with a yellow exclamation mark icon over it to show that the device has not been installed properly. Right click on this and choose Update Driver. Choose "No, not this time" from the first page and click next. Then choose "Install from a list or specific location (Advanced)" and click next again. Now click the "Include this location in the search" and click Browse. Navigate to the Drivers folder of the unzipped Arduino IDE and click Next. Windows will install the driver and you can then click the Finish button. The Arduino Uno will now appear under Ports in the device list and will show you the port number assigned to it (e.g. COM6). To open the IDE simply double-click the Arduino icon in its folder.

Installing on Windows 7 – Windows Vista

Once you have downloaded the latest IDE, unzip the file and open it. You will see the Arduino files and sub-folders inside. Next, plug in your Arduino using the USB cable and ensure that the green power LED (labeled PWR) turns on. Windows will attempt to automatically install the drivers for the Arduino Uno and it will fail. Click the Windows Start button and then click Control Panel. Now click System and Security, then click System, and then click Device Manager from the list on the left hand side. The Arduino will appear in the list as a device with a yellow exclamation mark icon over it to show that it has not been installed properly. Right click on the Arduino Uno and choose "Update Driver Software." Next, choose "Browse my computer for driver software" and on the next window click the Browse button. Navigate to the Drivers folder of the Arduino folder you unzipped earlier and then click OK and then Next. Windows will attempt to install the driver. A Windows Security box will open up and will state that "Windows can't verify the publisher of this driver software." Click "Install this driver software anyway." The Installing Driver Software window will now do its business. If all goes well, you will have another window saying "Windows has successfully updated your driver software. Finally click Close. To open the IDE double-click the Arduino icon in its folder.

Installing on MacOSX

Download the latest disk image (.dmg) file for the IDE. Open the .dmg file. Drag the Arduino icon over to the Applications folder and drop it in there. If are using an older Arduino, such as a Duemilanove, you will need to install the FTDI USB Serial Driver. Double-click the package icon and follow the instructions to do this. For the Uno and Mega 2560, there is no need to install any drivers.

Writing Arduino Software

So you have set up the Arduino development environment (IDE) and you are ready (and anxious) to write your first program.

The Arduino IDE comes with a C/C++ library called "Wiring" (from the project of the same name), which makes many common input/output operations much easier. Arduino programs are written in C/C++, although you don't really need to be very familiar with the latter. You need to be aware of some general programming principles (like variables, control statements, loops, etc.) and how programming functions (or methods) work. Many of the pre-installed Arduino example sketches are provided to also introduce you to the Arduino programming concept. Let's first see the basics of an Arduino sketch and then move on to some programming examples.

Figure 4-24. The Arduino IDE Environment where you can edit your sketch code and upload it to your Arduino board. You can also see the buttons for the main functions, like compiling and uploading sketch, opening the serial monitor and the message area.

The Arduino Sketch

When you open the Arduino IDE, you notice the main white area that waits from you some code. This is the main place that your Arduino sketch will reside when you edit it, open it from a source file and upload it on your board.

Starting with the Arduino sketch structure, you need to keep in mind that a sketch has two mandatory components:

- The **setup** function: This is identified in the sketch as void setup(). As the name probably suggests, it is the main initialization function. Whatever needs to be initiated or initialized with a value when the sketch will begin to run on your Arduino it shall be stated there. The setup function is run once and thus the code inside is executed only once (unless you have defined some repetition loops) and in the same sequence you have written it. Most usually you will find here the initialization of your variables and/or custom system settings (like initializing and setting the baud rate for the Serial

Monitor), the initialization of board pins as input or output pins, etc.

- You may call external functions from the setup() function. You can also refer to external variables (when defined globally, in the beginning of your sketch).

- The **loop** function: This is identified in the sketch as void loop(). This function is automatically executed (you do not have to invoke it manually in your code) after the execution of the setup() function. Anything contained in this function will be executed repetitively (as if it was in some while(true) {...} loop).

- This makes quite sense since you will be most probably using your Arduino to monitor some inputs or outputs and send data over the web continuously. In case you need to stop the execution of the whole loop code or a part of it you will need to insert that part of the code inside a control statement (if...then).

- In addition, you will find that in most cases there is a delay introduced in that function. It can be anywhere inside the loop code (depending on your project needs) and most usually you will also find it at the end of the loop function. This delay (usually set in microseconds) gives the option to the board to hold the execution of the program before it moves again to the beginning of the loop code or to the next code part.

Apart from the latter you may find in your sketch additional code components like variables, reference to external libraries, custom functions, etc.

Let's see some basic examples that will help you understand more the coding of Arduino sketches.

Some Basic Examples

Let's discuss some code examples that will demonstrate the basic features of the Arduino programming. These examples will give you the main knowledge you need to start writing your own sketches and also be able to comprehend the more complex code presented in the following Chapters.

Interfacing with the Serial Monitor

First we begin with the Serial Monitor. It can be considered as the most important feature of the IDE since it allows you to see some textual output that your Arduino is generating with the help of your code, which can be very useful especially when you try to debug your programs

The Serial Monitor (as part of the IDE) is actually an area where any message transmitted from your Microcontroller is displayed. Similarly, you

*To
a*

can send text to your Arduino board using the Serial Monitor. In both cases you need to define this communication in you sketch.

In order to be able to use the Serial Monitor you need to have two things in your sketch: a) The initialization of the Serial object with a specific communication baud rate (called baud rate), b) the use of essential Serial object functions that will enable you to either have data printed or read. Let's see the details in the following code listing:

Listing 4-1. Simple Example of Using the Serial Monitor

```
void setup() {
  Serial.begin(9600);
  Serial.println("Hello Arduino, ");
  Serial.print("I am a great fan!");
  Serial.println("I like Arduino, Sensors and the Cloud");
}

void loop() {
}
```

In the setup() function you initialize the Serial communication the 9600 baud rate. This is done using the begin() method. Then you start printing messages to the Serial Monitor. Notice the difference between the println() and print() functions.

Add this code into your editor and click the upload button (make sure your Arduino board is connected and you have selected the appropriate board and USB port from Tools menu). When upload is done, click on the Serial Monitor icon (or use Tools-> Serial Monitor from the Menu). The Serial Monitor window will pop up, the sketch will start again on the board (the code is executed on the board immediately after the sketch upload process has finished. In order to be able to monitor the outputs in the Serial Monitor from the beginning of your sketch, the execution is reset every time you open the Serial Monitor). On the low right corner of the window make sure that the selected baud rate is the same you have used in your setup() function (9600). You will see the following textual output:

```
Hello Arduino,
I am a great fan!I like Arduino, Sensors and the Cloud
```

Let's see now how you can send information to your Arduino through the Serial Monitor. Create a new sketch from scratch and add the following code:

Listing 4-2. Simple Example of Sending Data to the
Serial Monitor

```
void setup() {
```

```
    Serial.begin(9600);
}
char in;
void loop() {
    while(Serial.available()>0){
            in = Serial.read();
            Serial.print(in);
    }
}
```

Upload the sketch to your board and open the Serial Monitor when uploading is completed. On the upper part of the Serial Monitor window there is an editable textfield. Enter a message, anything, and hit enter. You will see the content of your message to be displayed in the monitor.

The code in the listing initializes the Serial monitor at the 9600 baud rate as in previous example. You also define a char variable (in) for storing the incoming data. Then in the loop() function you use the available() function of the Serial class. This method returns true in case data are received from the Serial (i.e. when you send your message). In such case the while loop is executed until no more data is received (when the messaged has ended). Inside the loop you store the incoming char data into the in variable, using the Serial.read() method. You also print each character that arrives so that you can see the output on the Serial Monitor.

Controlling I/O ports

In Cloud applications it is very common to control devices or reading input states using the I/O ports. The following code listings demonstrate how you can control the state of a LED by switching between HIGH and LOW states of a digital port and how to read the input of a digital port for getting external events like pushing a button.

Listing 4-3. Controlling a LED's state Through a Digital Port

```
void setup() {
  // initialize the digital pin as an output.
  // Pin 13 has an LED connected on most Arduino boards:
  pinMode(13, OUTPUT);
}

void loop() {
  digitalWrite(13, HIGH);    // set the LED on
  delay(1000);               // wait for a second
  digitalWrite(13, LOW);     // set the LED off
  delay(1000);               // wait for a second
}
```

Listing 4-4. Changing a LED's State Based on Button's Events

```
const int buttonPin = 2;      // the number of the pushbutton
                                 pin
const int ledPin =  13;       // the number of the LED pin

// variables will change:
int buttonState = 0;          // variable for reading the
                                 pushbutton status

void setup() {
  // initialize the LED pin as an output:
  pinMode(ledPin, OUTPUT);
  // initialize the pushbutton pin as an input:
  pinMode(buttonPin, INPUT);
}

void loop(){
  // read the state of the pushbutton value:
  buttonState = digitalRead(buttonPin);

  // check if the pushbutton is pressed.
  // if it is, the buttonState is HIGH:
  if (buttonState == HIGH) {
    // turn LED on:
    digitalWrite(ledPin, HIGH);
  }
  else {
    // turn LED off:
    digitalWrite(ledPin, LOW);
  }
}
```

In case you need more information and examples about Arduino sketches and projects you can check the Amazon site for Arduino books series that are suitable for both beginners and advanced Arduino users!

Trying your code on an Arduino Simulator

When playing with your Arduino and experimenting with features and capabilities you mind find yourself in the following situation: you have a nice idea about a new project, or you want to test an existing one, but you do not have the essential components. You might be missing LEDs, buttons, a potentiometer even an Ethernet shield (see Figure 4-27), but you still have your sketch ready. Seeing your sketch code compile in the Arduino IDE is not as satisfying as watching it work on your board and you also need to verify that your program logic works. What can you do without the hardware? You can use an Arduino Simulator like the VirtualBreadboard.

VirtualBreadboard (http://www.virtualbreadboard.net/) is a (unfortunately Windows only) simulation and development environment for Microcontrollers. It supports many of the PIC16 and PIC18 Microcontroller devices and the Arduino platform. In addition it provides a wide variety of simulated components such as LCD's, Servos, logic and other I/O devices that can be used to model and simulate high level circuits. By simulation environment we mean that you have access to virtual boards and components that you can virtually connect together through an appropriate graphical environment (see Figure 4-25). You can add program code (in a similar way you would compose a sketch), you can compile the code and watch it being executed on the Microcontroller platform and interact with the connected components (LEDs, sensors, actuators) the same way it would if your circuit was real!

Currently VirtualBreadboard supports all the basic sketch examples included in the Arduino IDE. Both the sketch code is available and the essential circuit components (already connected to a virtual Arduino board), like LEDs, buttons, LCDs, Servo motors, etc. to support the virtual execution of the sketches.

As illustrated in Figure 4-25 the main graphical environment of VirtualBreadboard provides you with a virtual representation of an Arduino board. On the left you can find a variety of components and circuit elements that you can add to your virtual workspace through drag-n-drop and connect them to the board's pins.

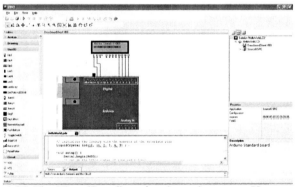

Figure 4-25. The VirtualBreadboard main environment. You can connect a variety of hardware to a virtual Arduino board and write the same sketch code you would with your real Arduino.

The main screen of the graphical interface is divided in two parts. The upper part contains the virtual representation of the circuit setup and lets you connect the devices together by also specifying what device pin goes to what Arduino pin, etc. The lower part contains the sketch editor. The editor has similar features with the Arduino IDE editor (syntax highlight, etc.).

Basically you can add exactly the same code (with some additional header information) you are using to your default IDE. VirtualBreadboard will compile the code and start (given that it finds no compilation errors) the virtual execution as in Figure 4-26.

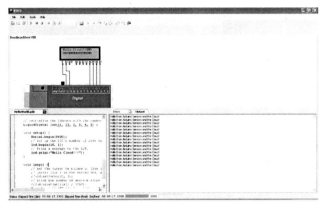

Figure 4-26. The VirtualBreadboard executes your sketch and you can see the results of your code both on the simulated Arduino board and the connected peripherals and the simulated Serial Monitor.

During the virtual execution you can see both the results of your code on the circuit you have created (e.g., LEDs turning on, servos moving, characters being print on LCDs, etc.), interact with it (e.g., turn potentiometer knobs, push buttons, etc.) and also see the output of the virtual serial communication with the Arduino board as if you were watching the Serial Monitor.

In addition to running your code virtually, the VirtualBreadboard can also upload the sketch directly to your real board when connected to your computer. So you can initially verify that the logic of your program is correct and then upload it for execution to a real board.

Extending Arduino using the Shields

Since Arduino has been introduced in 2005 it has been continuously evolving and enhancing. Not only has the official Arduino team been developing new boards, improving the IDE and adding more functionality, but also external vendors (like Sparkfun or Seeedstudio) have been creating extensions to the hardware itself. These extensions are knows as Arduino Shields.

Figure 4-27. The Arduino Ethernet Shield. It is designed to be plugged directly on top of Arduino boards, extend the functionality by providing the appropriate circuits and Ethernet socket, and at the same time allow user to access the I/O pins of the board.

Shields are circuit boards that can be plugged on top of an Arduino board extending its capabilities. Such capabilities can consider networking, like the Ethernet, the WiFly or the XBee shields. Figure 4-27 illustrates the Ethernet shield that features an Ethernet port. It can be plugged on top of an Uno or Mega board and at the same time expose the I/O pins of the board for usage. Shields can also be used to extend the ports of the board or allow users to connect sensors more easily, like the Electronic Brick Shield by Seeedstudio, they can add easy access to SD card memory modules, to Bluetooth and GPRS/3G modems and many more. The list of features that can be added to the Arduino through shields is really endless. Whatever new technology comes in the context of communication and interaction with circuits and electronics it is most likely that it can be implemented as a shield for the Arduino. The Arduino Shield List site (http://shieldlist.org/) currently lists over 250 Arduino shields with different functionalities!

The different shields always follow the same philosophy as the original Arduino: they must be easy to mount, and always provide users with appropriate libraries for using them.

Using Libraries with your Arduino

Apart from extending the Arduino hardware there are definitely many ways to extend the software as well by adding features that do not come with the default programming environment. Like in almost every programming language you can extend the abilities of your program by utilizing libraries, either existing ones (internal libraries) or external ones (that need to be imported to your programming environment).

In the case of Arduino libraries are files written in C or C++ (.c, .cpp) and provide your sketches with extra functionality (e.g. the ability to control an LCD screen or communicate with the Internet, etc.).

To use an existing library in a sketch simply go to the Sketch menu, choose "Import Library", and pick from the libraries available. This will insert an #include statement at the top of the sketch for each header (.h) file in the library's folder. These statements make the public functions and constants defined by the library available to your sketch. They also signal the Arduino environment to link that library's code with your sketch when it is compiled or uploaded.

In case you wish to use an external library you need to import it into the Arduino IDE. User-created libraries as of version 0017 go in a subdirectory of your default sketch directory. For example, on OSX, the new directory would be ~/Documents/Arduino/libraries/. On Windows, it would be My Documents\Arduino\libraries.

Import a Library to your sketch
Let's see an example of importing an external library to the Arduino environment and using it in your code.

In case you have not done yet so, go to the book code repository and download the source code for this book. Inside the Chapter 04 folder you will find a folder named "Arduino-Library-FancyLED". This is a user-contributed library that enables blinking LEDs custom times at custom rates. To import it, simply add this folder into the libraries directory of your default sketches directory. Then (re)start your Arduino IDE and go to File->Examples. You will notice there that the list with available examples has changed and now also includes a folder named 'FancyLed'. This folder contains 6 sketches that are the basic examples for using this library as provided by its creator.

Notice that this way you browse through the provided sketch examples for a particular library. To import the library in your own sketch do the following:

1. Create a new blank sketch (File->New)
2. Import the library by selecting Sketch->Import Library from the main menu. At the end of the existing libraries list you will see the 'FancyLed' library, under 'Contributed'. By selecting the library you will see that the following line is imported in your sketch code:

```
#include <FancyLED.h>
```

As in C/C++ programming, you include libraries using the #include statement.

3. Add the following code to your sketch to make some usage of the provided functions of the library:

```
FancyLED boardLED = FancyLED(13, HIGH);

void setup() {
}

void loop() {
  boardLED.turnOn();
  delay(1000);
  boardLED.turnOff();

}
```

As you might have guessed by simply looking the code, the sketch uses the library as an alternative way to blink the internal board LED (pin 13). It uses a FancyLED object that is initialized by the board pin number and the pin state. The object provides the methods turnOn() and turnOff() that set the state of the LED to On and Off respectively. The implementation of the code for the FancyLED class that enables as to construct objects of that type and use their functions is included inside the library you just have imported. You can check the source files in the "Arduino-Library-FancyLED" and see how the library is implemented.

This is a very simple example that illustrates how a library can hide the code implementation from the user and expose some specific code functionality by allowing users to create objects and use their methods.

For the main projects presented in this book that deal with Internet and Cloud-systems communication you will be using external libraries that will have to be imported in your Arduino IDE that same way you did with this simple example.

A great list with resources and information for (almost all) the available libraries that exist for the Arduino platform can be found in http://arduino.cc/playground/Main/LibraryList. The list is also continuously updated with new libraries and examples.

Running Arduino out of the Board

By definition, the Arduino board is a prototype board. You have a Microcontroller platform and environment at a considerable low cost and you can try easily and fast prototyping of your own projects. But what if you need to move ahead from the prototype and deploy a project more permanently or in larger scale? Do you need to buy yourself several Arduino boards?

The answer is fortunately no. You can easily program the Microcontroller chip (while being on board) and then remove it and use it as standalone in your circuits. Let's see a small example that will give you the basic idea and present you the materials you need in order to run Arduino sketches out of the board! The example is based on the Blink sketch. You will use an Arduino board like the Uno to program the Microcontroller with the sketch and then remove it and make it run on its own.

What you will need
You will need an ATMega Microcontroller (either ATMega128 or ATMega328) with Arduino Bootloader installed. You can either use the chip you will find on your Uno board or buy one. You will also need a pair of 22 pF ceramic capacitors and a 16 MHz crystal oscillator. Figure 4-28 illustrates the capacitors and the crystal next to the ATMega328 Microcontroller chip. The oscillator functions as an external clock – timer that indicates clock cycles to the processor. The ATMega328 comes already with an internal oscillator (at 8MHz) but it is better to demonstrate how to use an external one, needed especially in more complex projects.

In addition to the essential components so that the chip can run on its own, you will need power (you can power the chip easily from the Arduino board), a LED for this example, a Breadboard to set up your circuit and some jumper wires for connecting everything together.

As an alternative to the oscillator and the two capacitors you can get a ceramic resonator of same frequency (16MHz).

Some boards come with the SMD version of the Microcontroller. Obviously you cannot remove that chip and use it as standalone or it is quite hard to use one on a Breadboard. Make sure you get the 40-pin version of the chip as in Figure 4-28.

An example
Let's put everything together to run the Blink Arduino sketch on the Microcontroller on its own. First you need to program your Microcontroller chip. Assuming that it is already on the board, simply upload the sketch from the Arduino IDE. The sketch is located at Examples -> Basics -> Blink. You do not need to make any modifications to the sketch. Make sure you have selected the appropriate board from the hardware menu that reflects your Microcontroller type. Then just upload the sketch. When done, you will notice the internal board LED (defined as digital port 13) to blink.

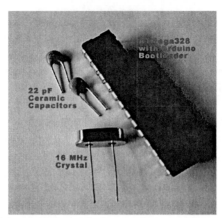

Figure 4-28. The basic materials for running an Arduino sketch on the Microcontroller itself (out of the box)

Let's get to the interesting part now. Remove first the USB cable from the board (make sure it is not powered by any other means). Then remove the Microcontroller chip from the socket of the board. You can use a small flat screwdriver to assist you in lifting it up. Be gentle while releasing its pins from the socket so that you don't bend any of them.

Now that you have released the chip you are ready to work with the breadboard setup. Based on the ATMega 168/328 Microcontrollers schematic (http://arduino.cc/en/Hacking/PinMapping168) you need to connect some of the pins to power and ground and connect the oscillator to the chip as well. More specifically:

- Pins 7, 21 and 20 need to be connected to a power source (Vcc)
- Pins 8 and 22 need to be connected to the Ground (Gnd)
- Pins 9 and 10 need to be connected to the crystal oscillator

The majority of the rest of the pins function as I/O pins. Based on your sketch and the Microcontroller schematic, you need to attach the positive pin of the LED (the longest pin) to pin 19, which corresponds to digital pin 13 for the Arduino.

For your convenience the appropriate connections as with the chip and the modules are illustrated in Figure 4-29.

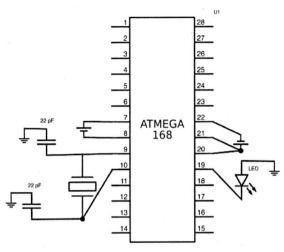

Figure 4-29. The schematic for connecting the capacitors, the crystal oscillator, the LED, Power (Vcc) and ground (Gnd) with the Microcontroller chip.

You can get the operating voltage for the Microcontroller chip and the ground connection you need from your Arduino board. First you need to place the chip in the center of your Breadboard as in Figure 4-30.

Programming your Arduino for the Internet of Things

So far you have seen what the Arduino is what it can perform (well you have seen in general the features and operational principles and hopefully you can imagine a lot of what it can do!). More project ideas and instructions for programming your Arduino board and circuit instructions can be found in numerous resources over the Internet and books.

This book is mostly dedicated to the communication of the Arduino with the Internet and Cloud applications. Such projects have special requirements in addition to reading inputs and controlling board outputs. The rest of the chapters in this book will give you more information on what you will need to communicate your Arduino with the Internet, your mobile phone and with Cloud-based services. The following sections will introduce you to special programming techniques that you might find useful when dealing with such projects. More specifically, we will discuss the use of timers in your sketches so that you can organize the way you read and transmit data over the web, you will see how you can parallelize your sketch execution through thread programming and finally how to make the transmission of your data over the web more secure using authentication and encryption.

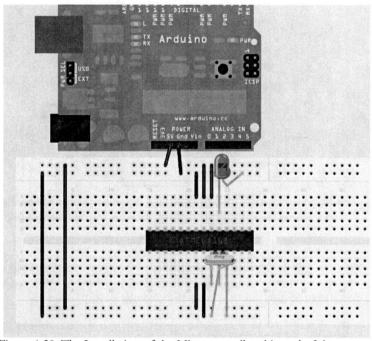

Figure 4-30. The Installation of the Microcontroller chip and of the rest components on the Breadboard. You can power your circuit directly through your Arduino board.

Figure 4-31. The ATMega328 on a Breadboard running the Blink sketch on its own!

Using Timers

When writing your own sketches for reading sensors, sending and receiving data and events from the Cloud, you will definitely face the usage of timers. Timers are functions that allow you to control when something particular shall happen in your program. Like when you need to read and transmit a sensor value over the Web. As previously discussed, the code defined in the loop() function is repetitive and is sequentially executed until you programmatically or externally (e.g., remove power or reset the board) stop it. If there is no delay introduced in the loop, the board will continuously do things like read and transmit sensor values. According to the ATMega Microcontroller specifications, the chip can read analog input values every 100us, which is about 10KHz or 10000 readings per second. If you try to print the sensor values to the Serial monitor the readings (although not at the same rate since additional delays are introduced) will be generated that fast that you can hardly read them while the board is still running the code. It can be also very hard to send data in such frequency over the Web due that most likely the receiving applications cannot handle such rates.

So it might become essential that introduce some delay or some timing function to your sketch in order to control the frequency of reading or sending data.

The default – native timing function in Arduino sketches is the delay() function. You will notice it in almost every sketch inside the loop() function and it is commonly used to give your board some time before it initiates the execution loop again. The function takes as input the number of milliseconds you wish to delay the execution for.

In many cases though you will need to add delays in several parts of your code to control different things. You might need to add a delay for initializing an Internet connection and two different delays for reading two different sensors. Instead of using the delay() function several times you can utilize Timer libraries that implement scheduling and delays in your program in a more efficient way.

There are several implementations of Timer libraries available. You can get an updated list from the official Arduino web site at http://arduino.cc/playground/Main/LibraryList. The most popular Timer libraries are the MsTimer2 and the Metro library. Let's see some examples!

The MsTimer2 Library

You can get the library from http://www.arduino.cc/playground/Main/MsTimer2. Import it into your Arduino environment by placing the files under on {Arduino-path}/hardware/libraries/. The default example that comes with the library is the one in Listing 4-1. It is an alternative implementation of the LED-blinking example but with the usage of the Timer instead of the delay() function.

Listing 4-1. Demonstration of the MsTimer2 Library

```
#include <MsTimer2.h>

void flash() {
  static boolean output = HIGH;

  digitalWrite(13, output);
  output = !output;
}

void setup() {
  pinMode(13, OUTPUT);

  MsTimer2::set(500, flash); // 500ms period
  MsTimer2::start();
}

void loop() {
}
```

The code starts by importing the library. Then a custom function is defined and implemented that contains the functionality for putting the LED on and off. The function uses a static boolean variable and changes its state every time it is invoked. Similarly the state of the LED changes.

Now the interesting part comes at the setup() function (and also at the loop() function which surprisingly is empty). An instance of the MsTimer is initialized with the interval time (500ms) and the function to be invoked every time the timer has expired (the previously defined flash() function). Then the timer instance is initiated. That is all it takes to start and use the timer. Since the program does not involve any other interactions, we do not need to make any implementation to the loop() function. The timer instance through its implementation in the library code, takes care of all the essential loop mechanisms and the time checking. This example has been quite simple, but you can imagine how assistive the library can be in case you need to schedule more than two events (e.g., reading from a sensor and moving a servo motor) at different time intervals. Instead of messing with control statements and using the delay() function several times, you can just define different functions for your events and then initialize the appropriate number of MsTimer2 instances.

The Metro Library
The Metro library is another timer library that facilitates the implementation of recurring timed events like blinking LEDs, servo motor control and

serial communication. You can download the library from http://www.arduino.cc/playground/Code/Metro.

The Metro library follows a different philosophy regarding usage compared to the MsTimer2 one. Let's see the following code example.

Listing 4-2. Demonstration of the Metro Timer Library

```
#include <Metro.h>
#define LED 13

int state = HIGH;

Metro ledMetro = Metro(250);

void setup()
{
  pinMode(LED,OUTPUT);
  digitalWrite(LED,state);
}

void loop()
{

  if (ledMetro.check() == 1) {
    if (state==HIGH)   {
      state=LOW;
      ledMetro.interval(250);
    }
    else {
      ledMetro.interval(1000);
      state=HIGH;
    }
    digitalWrite(LED,state);
  }
}
```

The code above will blink a LED attached to pin 13 on and off. The LED will stay initially on for 0.25 seconds and then stay off for 1 second.

You start as usually by importing the variable and by defining the board LED pin (pin 13). You also create a variable to hold the LED's current state. You instantiate a metro object and set the interval to 250 milliseconds. In the setup() function you set the pin mode to output (so that you can generate the HIGH and LOW states) and put the LED on HIGH state at the beginning. Inside the loop() function you check if the metro has passed it's interval through invoking the library's check() method. If the interval has expired the method will return true (1). In such case and if the pin is HIGH, you set the interval to 0.25 seconds. If the pin is LOW then you set the interval to 1 second.

The main difference with the MsTimer2 library is that you use the Metro check() method to verify if the timer has expired inside the loop() function.

Using Threads

Based on your experience with C/C++ programming language you might have noticed that so far the Arduino sketches are executed sequentially. The latter means that when you add a delay in your sketch after reading a sensor the whole execution will be stalled before the Arduino can move on with doing something else like sending data over the Internet or reading a second sensor.

One solution to address this issue is to use an external Thread library like the mthread. You can download the library files from here: https://github.com/jlamothe/mthread. In order to use the library you also need to also download and import the newdel library from here: https://github.com/jlamothe/newdel

The mthread library provides a Thread class that can allow you to execute functions in parallel inside your Arduino sketch. This class can be derived from to provide the basic functionality for a single thread. Based on the implementation of the library, each thread that is created through the library has its own loop() function. The latter works in basically the same way as the loop() function in your main Arduino sketch. The only major difference is this function in a Thread returns a boolean value (true if the loop needs to be called again or false if it doesn't. When the Thread completes, it will automatically destroy itself freeing the memory it occupied to be used for other things.

Let's see the following code example that utilizes Threads.

Listing 4-3. Using Threads for Parallel Execution Utilizing the Mthread Library

```
#include <newdel.h>
#include <mthread.h>

class FooThread : public Thread
{
public:
  FooThread(int id);
protected:
  bool loop();
private:
  int id;
};

FooThread::FooThread(int id)
{
  // Set the ID:
  this->id = id;
```

```
}

bool FooThread::loop()
{

  // Die if requested:
  if(kill_flag)
    return false;

  // Print the status message:
  Serial.print("FooThread ");
  Serial.print(id);
  Serial.println(" called.");

  sleep(1);
  return true;

}

void setup()
{

  for(int i = 1; i <= 5; i++)
  {
      main_thread_list->add_thread(new FooThread(i));
  }

  Serial.begin(9600);
  delay(1000);
}
```

As usually you import the essential libraries in the beginning of the sketch. Then you create a custom thread class that extends the Thread class of the library. Inside the class you define the constructor that takes an id as input parameter (which will help us to identify which thread is executed at runtime) and you also define the loop function (the one that belongs to the thread).

You move on by implementing the constructor function of the class and the loop function. Inside the loop function you check for a particular flag (called kill_flag) that will indicate the termination of the thread. You print some messages to the Serial Monitor and you delay the execution for 1 second. Note the difference between the delay() function and the sleep(). Depending on the id of the thread you could alternatively implement here the code that would read values from different sensors on your circuit.

On the setup() function you construct and initiate 5 threads using the class you have just implemented (new FooThread(i)). In order the threads to be initiated you add them to the main program thread, called 'main_thread_list' through the add_thread() function. Finally, you initialize the Serial Monitor so that you can watch the output of the threads and you wait 1 second before you allow the threads to be executed (remember that the delay() function holds the execution of the entire program).

When execute the code above will generate the following in the Serial Monitor:

```
. . .
FooThread 1
called
FooThread 2
called
FooThread 3
called
FooThread 1
called
FooThread 4
called
FooThread 2
called
. . .
```

You can see that while the threads are being created they are also being executed in a parallel manner.

Adding Security to your Sensor Readings

The basic concept behind the Internet of Things is the communication and data exchange of sensors and other embedded devices with the Internet. This introduces several issues, like how to perform the data communication, energy issues and of course the issue of data security. Data security in such web-based applications is translated in achieving two tasks: proper authentication and data encryption. The first considers mechanisms (like using a username and a password or a secret key) so that your web service for example can identify your Arduino board so that no other user (or no other board) can send data to your service. Encryption in other words means applying cryptography to your data. Cryptography is math oriented and uses patterns and algorithms to encrypt the data like messages, text, words, signals and any other forms of communication you might want to achieve between your sensors and a Cloud service.

Usually, two forms of encryption are used nowadays: symmetric and asymmetric. In symmetric-key encryption, each computer has a secret key (code) that it can use to encrypt a packet of information before it is sent over the network to another computer. Symmetric-key requires that you know which computers will be talking to each other so you can install the key on each one. Symmetric-key encryption is essentially the same as a secret code that each of the two computers must know in order to decode the information. The code provides the key to decoding the message.

The other type of encryption is through the use of asymmetric key algorithm, which uses both public-key and private-key cryptography. With this method, a user can send data via the public-key that is then encrypted, while the receiver, who is only one who can decrypt the information, uses the private-key. However this kind of cryptography requires a lot of processing that the Arduino cannot handle. For simple projects like the ones presented in this book it is still safe to use basic symmetric encryption using the same key both in the Arduino code and the decoding application.

In Cloud-based systems and IoT platforms there are mainly two reasons you might consider essential the authentication and data encryption: a) your sensor information can be too sensitive to store online as plain textual information that one could have access to, b) the information is always transmitted over the Internet using often untrusted networks. In addition, Arduino cannot transmit information over SSL (HTTPS) protocol, which suggests that the data can often be transmitted in clear textual format. Thus, it can be quite useful in some cases that you utilize both authentication and data encryption. Note that you have to implement the web-based services or the applications that receive data from the Arduino accordingly, so that they support the authentication and/or encryption mechanisms you use.

For your Arduino projects you will be presented with libraries and sketches that allow you to do both authentication and data encryption.

Authenticating your Arduino

The first part of the Arduino security involves the proper authentication of your Arduino so that you can be sure your application receives data only from your board. The idea is adopted from the common web-based authentication systems that require username and password form users. To such systems, the latter credentials are usually stored into application databases so that they can be compared with what users provide upon authentication. However this user information is not stored as plain text since it introduces great security risks of being compromised. What is done in most cases is that a hashing mechanism is used that encodes the user authentication data. The same encoded data (with the same exact method) exist in the authentication application. Comparing the latter can give us a safe indication whether the credentials are correct or not.

(Cryptographic) Hash functions are methods that take as input a string (or an object) and return a fixed-size string that is unique and strongly related to the input. The latte means two things: a) there is no way (more correctly it is highly improbable) to get the same output for two different inputs, b) even the slightest change in the input data produces a totally different output. In addition there is no way from the encoded message to retrieve the original one. This makes hash functions suitable for use in authentication since they can guarantee that even if someone gets the encoded information, there is no way to reveal the actual data. Hash functions work similar with the checksum functions that generate a code of fixed size (checksum) for a particular input and are used mostly to verify the integrity of file data during operations like network transmission, media storage, etc.

Several Hash functions exist and are used for such purposes. Most popular are the MD5 and SHA1/SHA256 functions.

For your Cloud-based projects it might be useful to utilize such Hash functions to authenticate communication between your Arduino and a web-based application. Currently, most sensor-related Cloud applications (as you will also explore in Chapters 8 and 9) use only authentication keys transmitted as plain text, but you can build your own Cloud application that will support Has-based authentication with your Arduino (as in Chapter 7).

Encode Key/Username with SHA1 Hash function

For the following Arduino sketch you utilize the 'Cryptosuite' library for Arduino. This library provides Arduino-compatible implementation and the appropriate functions for the SHA1 and SHA256 Hash functions. You need to download the library files from here http://code.google.com/p/cryptosuite and install them on your Arduino environment as previously instructed.

Run the following sketch to your Arduino and open the Serial Monitor. Check the output and you will see that the hash-encoded output of the original message will be "7c02a8797b4a101e242c095a32faad0d84d82ff4". You can try with different strings as input and then verify the output with a web-based SHA1 generator like in http://www.webtoolkit.info/demo/javascript-sha-1/

Listing 4-4. Arduino Code Sample using SHA1 Hash Function to Encode a Key or a Username for Authenticating your Arduino

```
#include "sha1.h"
//The secret key (username) to store the encoded output by
the hash function
uint8_t* sectretKey;

//Help function for printing the output of the hash function
void printHash(uint8_t* hash) {
  int i;
  for (i=0; i<20; i++) {
    Serial.print("0123456789abcdef"[hash[i]>>4]);
    Serial.print("0123456789abcdef"[hash[i]&0xf]);
  }
  Serial.println();
}

void setup() {
  uint8_t* hash;
```

```
Serial.begin(9600);

//Initialize the SHA1 function
Sha1.init();

//Pass the username or sensor key to function
Sha1.print("MyArduinoSectretName");

//Get the hash of the usernsame or sensor key
sectretKey = Sha1.result();

//print the key in the Serial monitor
printHash(sectretKey);

}

void loop() {
  //here you can read some sensor data
  //...
  //before sending data to the Cloud
  //you can use the sectretKey to authenticate
}
```

Encode Key/Username with SHA1 Hash function – Code Review

You start your sketch by including the essentials from the 'Cryptosuite' library. In this example you utilize the SHA1 Hash functions so you need to include the respective header file only.

```
#include "sha1.h"
```

You define a variable for the secret key (e.g., a username) to store the encoded output by the Hash function. The unsigned integer type (uint8_t) is required by the library function. Since the key will be alphanumeric (i.e. sequence of both numbers and letters) you need to store a number of integers within the variable's memory so you use a pointer.

```
uint8_t* sectretKey;
```

You implement here a help function for printing the output of the Hash function. Same as the input, the output is in unsigned integer format. The function will check for integers in the input's variable memory and convert them to numbers and characters and print them to the Serial Monitor.

```
void printHash(uint8_t* hash) {
  int i;
  for (i=0; i<20; i++) {
    Serial.print("0123456789abcdef"[hash[i]>>4]);
    Serial.print("0123456789abcdef"[hash[i]&0xf]);
  }
```

```
    Serial.println();
}
```

You move one with the setup() function that initializes your sketch. You setup the Serial to the appropriate rate so you can watch the output on your Serial Monitor.

```
void setup() {

Serial.begin(9600);
```

You initialize the SHA1 function:

```
Sha1.init();
```

You then pass the username or sensor key to be coded as an input parameter to the function:

```
Sha1.print("MyArduinoSectretName");
```

You store the coded output (known also as hash from the name of the method) to the uint8_t variable you have previously defined:

```
sectretKey = Sha1.result();
```

Finally, you can print the output in the Serial monitor and see how it looks like. You can also try to modify (even change a letter from capital to small) and see how different the hash output becomes!

```
printHash(sectretKey);
```

You have not implemented anything specific inside the loop() function. Usually you will need to authenticate your Arduino once, either in the beginning of the sketch execution (i.e. in the setup() function) or through using a flag inside the loop() function.

For example, you define a Boolean flag variable somewhere in the beginning of your sketch:

```
boolean once = true;
...
void loop() {
    if(once) {
        //authenticate
        once = false;
    }
}
```

Then you can read some sensor data and forward them to the Internet. More information on how to do this on the following chapters!

Encrypting your Data

For data encryption you will use the ArduinoAES256 library. This library contains an implementation of the AES symmetric encryption algorithm specially ported for the Arduino platform. As the name suggests, you will need a symmetric key for encrypting your data. The same key will be used also in the decryption application so that the original message can be revealed. For this project you will use the following sequence of alphanumeric characters as a key:

```
000102030405060708090a0b0c0d0e0f101112131415161718191
a1b1c1d1e1f
```

You will also demonstrate the effect of encryption on a message and how it can be decrypted again using the same key. You can get the encryption library from https://github.com/qistoph/ArduinoAES256 and install it into you Arduino environment. Then go and try the following code!

Arduino code with AES Encryption

Listing 4-5. Arduino Code Sample using AES to Encrypt and Decrypt a Message

```
#include "aes256.h"

//Define the AES context
aes256_context ctxt;

//In the setup function you need to
//initialize the AES algorithm providing the
//encryption key
void setup() {
   Serial.begin(9600);

   Serial.print("Initializing AES256... ");
   //Let's use the following alphanumeric sequence (as
integers) as encryption key
   uint8_t key[] = {
      0x00, 0x01, 0x02, 0x03, 0x04, 0x05, 0x06, 0x07,
      0x08, 0x09, 0x0a, 0x0b, 0x0c, 0x0d, 0x0e, 0x0f,
      0x10, 0x11, 0x12, 0x13, 0x14, 0x15, 0x16, 0x17,
      0x18, 0x19, 0x1a, 0x1b, 0x1c, 0x1d, 0x1e, 0x1f
   };

   //You initialize the AES algorithm with the key
   aes256_init(&ctxt, key);
   Serial.println("done");

   char *message  = "Arduino, Sensors and the Cloud";
```

```
aes256_encrypt_ecb(&ctxt, (uint8_t*)message);

Serial.println("encoded message: ");
Serial.println(message);

aes256_decrypt_ecb(&ctxt, (uint8_t*)message);
Serial.println("original message: ");
Serial.println(message);

aes256_done(&ctxt);
}

void loop() {
}
```

Arduino code with AES Encryption – Code Review

You start the sketch by including the appropriate library, "aes256.h".

```
#include "aes256.h"
```

Then you define an essential object needed by the library, the aes256_context:

```
aes256_context ctxt;
```

The context if used in order to keep information like the encryption key and initializes the encryption mechanism. The initialization will be performed within the setup() method. You also provide the symmetric key there as an array of integers. The same key has to be used (and the same method of course) at the decryption application.

```
void setup()
```

You will use the Serial for monitoring the outputs of the encryption method:

```
Serial.begin(9600);
```

The sequence of integers as encryption key

```
uint8_t key[] = {
    0x00, 0x01, 0x02, 0x03, 0x04, 0x05, 0x06, 0x07,
    0x08, 0x09, 0x0a, 0x0b, 0x0c, 0x0d, 0x0e, 0x0f,
    0x10, 0x11, 0x12, 0x13, 0x14, 0x15, 0x16, 0x17,
    0x18, 0x19, 0x1a, 0x1b, 0x1c, 0x1d, 0x1e, 0x1f
};
```

We then initialize the AES algorithm with the key:

```
aes256_init(&ctxt, key);
```

Let yourself know that the Arduino are done with the initialization process.

```
Serial.println("done");
```

You move on with the encryption of the message "Arduino, Sensors and the Cloud". The AES256 library provides all the essential methods for encrypting and decrypting data. The default input of the encryption (and the decryption function as you see later on) is the encryption context (that contains the symmetric key) and the message to encrypt as a pointer to an unassigned integer. Since you want to encrypt a string message as a sequence of char, you need to perform the appropriate casting to uint8_t*.

```
char *message  = "Arduino, Sensors and the Cloud";
aes256_encrypt_ecb(&ctxt, (uint8_t*)message);
```

Let us know that the encryption is completed and also print the encrypted message on the Serial monitor so that you can see what it looks like.

```
Serial.println("encoded message: ");
Serial.println(message);
```

Now you move on to decrypt the message in order to demonstrate the reverse function and verify that you can receive the original message.

```
aes256_decrypt_ecb(&ctxt, (uint8_t*)message);
Serial.println("original message: ");
Serial.println(message);
```

Finally, you use the _done() method to clear the memory from the encryption context.

```
aes256_done(&ctxt);
```

In this project the loop function does not perform anything particular. You can use it however to acquire sensor readings and encrypt them before your send them to a Cloud application. More details on how to read from sensors and communicate your Arduino with the Internet on the following chapters!

Summary

This chapter has introduced you to the basic concepts of Microcontrollers, the brains of every embedded device. You have been briefly presented to the main components, to what Microcontroller programming is and why the use of a Bootloader makes embedded programming much easier! The Arduino Bootloader is thus one of the great features of the Arduino platform but not the only one. Arduino comes in many board versions and variations that can meet almost any kind of project requirement you might have. Even if you have some special requirements like wired or wireless communication, there might be an extension shield available that can provide you the required functionality.

Regarding programming the Arduino you have seen the basics of the Arduino IDE and the essentials for creating and running your own sketches. The IDE provides a series of sketch examples that demonstrate the programming and the abilities of the Arduino. In addition there is a great number of external libraries available that you can import in your IDE and enhance the features of your sketches. There are even particular libraries that you can use especially when programming for the Internet of Things, like timers, threads and authentication/encryption libraries. In case you lack some hardware or circuit electronics you can always use the VirtualBreadboard simulator and see your code running on a virtual Arduino board and interact with virtual electronic components.

So now you know precisely what the Arduino platform is and how you can program it. Let's move on to the next chapter and see in more details how you can sense your environment with the help of Arduino and analog or digital sensors.

5 READING FROM SENSORS

The previous Chapter has demonstrated you the basics of the Arduino platform, as well as has presented you some advanced issues like running Arduino of the board, using threads, using data encryption and other external libraries that might become useful when communicating with Cloud applications.

This Chapter will take you one step further and show you how to use your Arduino board and connect it to various sensors using the most common communication protocols. More specifically, we will discuss how to connect your board to sensors using the available interfaces, provide examples of how to read analog sensors and also cover all the variations of digital sensor interfaces that are available (digital ports, Rx/Tx ports, I2C ports and SPI ports). You will be also demonstrated how to read sensor data from another Arduino board, serving as an extension to your current one.

Sensing the World

You have been presented with various sensors in Chapter 2 and you must be eager to know how to use them. As discussed so far, the Arduino platform enables you to retrieve readings from both digital and analog sensors. According to the board type used (Uno, Mega, Mini, etc.) the number of available analog and digital ports varies.

Bur how does the Arduino board sense the world? In case of digital sensors that produce various HIGH and LOW states in their output (or sequence of them representing bits and bytes) things are pretty clear, since the Microcontroller can read two input values (0V for LOW and 5V for HIGH). In case of analog sensors (that generate an output voltage varying from 0V to 5V) an extra component is used in order to convert analog values to digital ones. This component is the analog-to-digital converter (ADC) that reads the changing input voltage and converts it to a binary value that the microcontroller can interpret. The ADCs are usually built into microcontrollers.

As you already know, the Arduino boards are based on the ATMega microcontroller family. The latter integrate a varying number of ADCs, therefore the number of analog ports you may find on an Arduino board can be between 6 (in case of the Uno) and 16 (in case of the Mega). The analog ports can however be used also as digital ports (in case all the existing digital pins have been used). Another great feature of the Arduino platform is that the analog pins integrate a pull-up resistor internally that protects the board from shortcuts and invalid sensor outputs (more details on pull-up resistors in Chapter 2).

The conversion of the analog signal to digital is made based on the sampling theory presenting in Chapter 2. Each Arduino board comes with a crystal oscillator (remember that component in previous chapter where you needed it to run the microcontroller of the board) that helps the microcontroller to count time intervals. The 16MHz oscillator on the Arduino boards means that the microcontroller can sample the analog signal for digital conversion about 16 x 106 (million) times per second.

When the microcontrollers senses a varying voltage, it converts that voltage to a number of a specific range. The range depends on the so-called 'resolution' of the microcontroller. The Arduino family has 10-bit resolution which means that the range for reading analog sensors is between 0 and 1023 (1024 values in total as a product of 210).

Reading from Analog Sensors

The placement of the analog sensor ports depends on the board type you are using. On the Arduino Uno they are located on the lower right corner of the board (as in Figure 5-1), named as A0-A5.

Figure 5-1. The 6 analog ports of the Arduino Uno board

The best way to connect a sensor is to use a breadboard and a jumper wire. Jumper wires feed pretty well in the Arduino ports and can save you a lot of time and effort from cutting simple wires and soldering parts together.

Usually an analog sensor will need to be connected to a power source and also grounded so that it can form a circuit. Analog sensors usually come in two variations with respect to their output pins. They either have only two pins (like resistors) or three pins for connecting to power, ground and providing the analog output (go back to Chapter 2 and see some examples).

Most Arduino boards provide two voltage outputs, a 5V and a 3.3V one and usually three Ground (annotated as GND) connections.

You need to be careful with your sensors operating voltage. Providing 5V to a 3.3V-operating sensor can damage it!!

When analog sensors come with 3 connection pins, it is more convenient, since they can be directly connected to your Arduino board. Let's see such an example with the TMP36 temperature sensor illustrated in Figure 5-2.

Figure 5-2. The TMP36 temperature sensor with 3 connection pins (image courtesy of Sparkfun)

According to the datasheet of the sensor (it can be retrieved from here: http://www.sparkfun.com/products/10988) the first pin (from left to right) is connected to power, the second provides the analog output of the sensor and the third one is connected to the Ground. You can connect directly the sensor with your Arduino board as in Figure 5-3. The analog output pin of the sensor is connected to the analog pin 0 (A0) of the board.

The sensor will get power from the board (according to the datasheet it can operate between 2.5V and 5.5V so you can power it either with 3.3V or 5V from the board) and will output a voltage variation on A0 according to the sensed temperature.

Reading from the analog port is quite simple in the Arduino sketch. All you need to do is use the analogRead() built-in function, defining the sensor pin (A0 in this case) and saving the output on a variable for further usage:

```
int reading = analogRead(A0);
```

Note that you can use both A0, A1, A2… and 0, 1, 2,… for declaring an analog pin number within your code.

As we mentioned before, the variation measured on A0 will be between 0 and 1023. Upload the following sketch on your Arduino, open the Serial Monitor for watch for the output:

Listing 5-1. A Simple Sketch For Printing The Reading of Analog Sensor Pin A0 to the Serial Monitor

```
void setup() {
  Serial.begin(9600);
}

void loop() {
  int reading = analogRead(A0);
  Serial.println(reading );
delay(500);
}
```

You will notice a variation of values according to the temperature sensed.

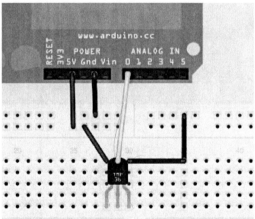

Figure 5-3. Connecting the TMP36 temperature sensor to your Arduino with the help of a breadboard

In order to convert those values in temperature you need first to convert the 0-1203 range to voltage. To do so you multiply the sensor reading by 5 or by 3.3 (according to the voltage you have provided to the sensor through the board, in the figure above it is connected to 5V) and divide the product by 1024:

```
float voltage = reading * 5.0;
voltage /= 1024.0;
```

According to the datasheet of the temperature sensor, each degree temperature change corresponds to a change of 10mV with a 500mV offset. So in order to get Celsius out of the voltage reading you need to subtract the 0.5 offset and multiply the result by 100:

```
float temperature = (voltage - 0.5) * 100 ;
```

Let's see now a different example with an analog sensor that has two connection pins. For this case you will use a light dependent resistor (LDR) to sense the light conditions. As the name suggests, it is a resistor and thus it is connected in a circuit like a resistor does, has two pins and no polarity.

Connect one pin of the LDR to the 5V output of your Arduino and the other pin to an analog input port, e.g., port A1 as in Figure 5-4. To make a proper circuit and let electric current flow through the LDR you need to also connect it to the Ground. To do so you add a pull-down resistor of 10K Ohm between the leg of the LDR that goes to the A1 port and the Gnd port of the Arduino board.

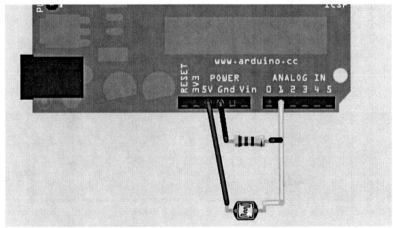

Figure 5-4. Connecting the LDR sensor directly to your Arduino using a pull-down resistor

To measure the variations of the LDR according to the lighting conditions in your room, simply use the code of Listing 5-1, but remember to change the analog input port to what you have connected your LDR to (A1 in case of the circuit in Figure 5-4). To turn the reading of this circuit into illumination values (called lux) you need to advise the datasheet of the LDR about the conversion formula from resistance to lux values.

Digital Sensors

When you will want to build your own project you will probably start with looking for available sensors to use in your circuit. You will notice that the majority of them are digital ones, and even the analog sensors come also in a version with a digital output. How come there are so many digital sensors available when they are supposed to measure a physical phenomenon in a range of values, like a physical signal (like temperature, humidity, distance, pressure, etc.)?

The answer is that digital sensors have integrated circuits that do plenty of the math for you and provide direct information about the measuring quantity (e.g., Celsius instead of voltage) and in many cases you will find that they can combine readings into one components (e.g., measure and output both humidity and temperature readings). They are also more precise in readings and less tolerant to errors. In addition, they can contain useful information (like calibration readings) in internal memory that you can use in your code in order to properly measure a phenomenon.

Digital sensors are falling into several sub-categories, not based on the number of connection pins, but based on the communication protocol they are utilizing.

At first, let's see how they can be connected on the Arduino board. Figure 5-5 shows the upper part of an Arduino Uno board that contains the 14 available digital ports. As mentioned before, the number of available digital ports and their placement depends on the board type you are using. For instance, the Arduino Mega 2650 board has 54 digital pins.

Figure 5-5. The 14 digital ports on an Arduino Uno board. Notice that the first two of them are used for Serial communication.

The digital sensors can be categorized into those that have on/off states like buttons, those that use the Serial Protocol for communication with the board, those that use the I2C protocol and finally, those that communicate over the SPI protocol.

Let's see in the following sections some examples of each category.

Sensors With On/Off Sates
The first category refers to sensors like push buttons or switches that produce an On/Off state. It may sound unnecessary to log button events on the Cloud, but you might want to monitor how often your outdoor is opened and when.

To demonstrate how to build such an event detection circuit you will see the function of a Reed switch and present you how to read the on/off states of it.

A Reed switch is a switch that is activated when it is exposed to a magnetic field. Think of the door alarm systems that have a mobile part attached on the door (a magnet) and a steady module attached on the door frame. The latter is simply a Reed switch that tells the alarm system when the door is opened or closed.

For this example you can also use a simple switch or an on/off button as alternatives to the Reed switch. Connect them as in Figure 5-6 and make sure you also use a resistor (220 Ohm value or similar will do fine) between the switch and the Gnd connection (As mentioned Arduino has internal pull-up resistors that will also protect it from shortcutting the 5V pin to the Gnd when the switch is On, but still it is a good practice to use resistors in any case).

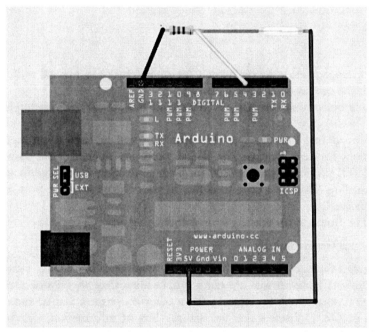

Figure 5-6. The circuit setup for reading the states of a Reed switch

Upload the following sketch to your board and open the Serial Monitor.

Listing 5-2. Reading the States of A Reed Switch and Printing them On the Serial Monitor

```
// The pin the reed switch is attached to
int reedswitch = 4;

// A variable for reading the switch state
int switchState = 0;

void setup() {
  Serial.begin(9600);
  // Initialize the reed switch pin as an input:
  pinMode(reedswitch, INPUT);
```

```
}

void loop(){
  // Read the state of the switch
  switchState = digitalRead(reedswitch);

  // Check if the state is HIGH or LOW (switch
activated/deactivated)
  if (switchState == HIGH) {
    Serial.println("Switch is activated!");
  }
  else {
    Serial.println("Switch is deactivated!");
  }
}
```

Take a magnet close to the switch (or open/close the manual switch) and watch the output on the Serial Monitor.

You will notice that the main differences of this sketch with Listing 5-1 are two:

4. You define that the specific digital port will be used as an input. This happens because the digital ports are general input-output (I/O) ports and can be used for both generating digital signals and reading. Thus you need to define in your sketch how the port will be used.

5. The function for reading from a digital port is:

digitalRead(reedswitch);

The rest of the code is much similar to when reading analog sensor values. On/Off State-sensors are quite simple since they trigger one event at the digital port. When it comes to more complex sensors that produce a sequence of On/Off states that are actually bits of information, sketches require some additional coding, as you will notice in the following examples.

Using the Serial Protocol

The Serial Protocol (often referred also as UART or USART) was one of the initial protocols for communication between devices and computers. The Arduino uses it in order to communicate with your computer during a program execution for both sending and receiving data (whenever you use the Serial monitor) and also for receiving a new sketch from the Arduino IDE. The communication is performed over digital pins 0 (RX) and 1 (TX) as well as via USB. (The Arduino Mega board has additional combinations of Rx/Tx pins).

The Serial Protocol is mostly used to connect devices like RFID and GPS. Let's see an example of reading an RFID sensor.

You can get a serial RFID reader from Parallax or from Seeedstudio (look for the 125KHz RFID UART module).

The following code works for the Parallax module, but you should be able to easily modify it and make it work for the Seeedstudio one. Upload the code first before connecting the RFID module to the Rx/Tx pins of the Arduino board. Since the sketch uploading via the USB is using the Serial protocol, there might be cases when the Rx/Tx pins on the board are occupied and the sketch upload fails.

Listing 5-3. Reading RFID Tags Using The Serial Protocol

```
// variable for defining the ASCII header character that
precedes each tag
int header = 10;

// variable for defining the ASCII carriage return
terminates each tag
int endByte = 13;

// the length of each tag
int tagLength = 10;

// variable for saving the tag ID as string
String tag = "";
int bytecounter = 0;

void setup(){

  // Needs to be set at the baud rate of your RFID reager
  Serial.begin(2400);

  // set pin 8 as an output pin
  pinMode(8,OUTPUT);
  // send a LOW signal to the pin in order to enable the
RFID reader
  digitalWrite(8, LOW);
}

void loop(){
  // check if the reader has sent at least 12 bytes (the tag
length + 2 bytes: 1 byte as header and 1 byte at the end
  if(Serial.available() >= tagLength + 2) {
    //Chech if the reader has sent the header
    if(Serial.read() == header){
      bytecounter = 0;

      // read the 10 digit code
      while(bytecounter < tagLength) {
```

```
    int val = Serial.read();
    // check if we have reached the end of the tag code
    if((val == header)||(val == endByte))
      break;
    //Otherwise add the bytes into the tag variale
    tag += val;
    bytecounter++;
  }
  // check for the correct end character
  if( Serial.read() == endByte) {
    //Get the tag id by removing the last character
    String tagID = tag.substring(0, tag.length()-1);
  }
  }
 }
}
```

A tag consists of a start character (like a header) followed by a 10-digit id and is terminated by an end character. This means that the reader transmits data in blocks of 12 bytes in total. Therefore the sketch above will first activate the RFID reader in the setup() function using the following line:

digitalWrite(8, LOW);

and then will start checking for receiving a sequence of 12 bytes in the loop() function. When that happens it checks for the header and the end character and add the incoming bytes into a String variable that is trimmed finally (removing the last character) to the RFID ID number.

In order to print the ID and read it in your Serial Monitor you can use the SoftwareSerial Library as explained in the following Section.

To build the circuit as in Figure 5-7, connect the Rx/Tx pins of your Arduino board with the ones from the RFID (RFID Tx pin goes to Arduino Rx pin and RFID Rx pin goes to Arduino Tx pin respectively) and the Vcc pin with the 5V pin of your Arduino and the Gnd pin with the Gnd pin of your board as well.

In the case of the Parallax RFID reader you got only a Serial out pin that goes to your Arduino's Rx pin, like in the following figure. You also need to connect the ENABLE pin (second pin from left to right) to a digital pin of your Arduino (e.g., pin 8). This serves as an activation switch to the module and requires a LOW state from the pin in order to activate the RFID reader.

The SoftwareSerial Library

As mentioned previously the Arduino hardware has built-in support for serial communication using digital pins 0 and 1. What happens if the latter pins are already occupied (e.g., you need to send data to your computer) and you still need to communicate with a device using the Serial Protocol, like the RFID Reader of the previous example?

The answer to the question is the SoftwareSerial library. The latter has been developed to allow serial communication on other digital pins of the Arduino, using software to replicate the functionality of the serial ports. According to the Arduino team, it is possible to have multiple software serial ports with speeds up to 115200 bps.

To use the library in your sketch, all you need is to include it in the beginning of it (the usual way you import external libraries) and define the additional digital pins that will serve as a serial port (you need a pair of Rx and Tx pins for each serial port you define).

The SoftwareSerial library is already included in your Arduino IDE. You might however consider using an improved version of it, called NewSoftSerial Library, provided by Mikal Hart at http://arduiniana.org/libraries/newsoftserial. The usage is identical but the latter is quite faster in receiving events and data from defined serial port.

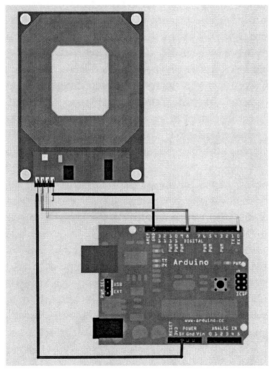

Figure 5-7. Connecting the Parallax Serial RFID Reader to your Arduino Board

Utilizing the SoftwareSerial library in the RFID reader sketch can allow you to use the default Arduino serial mechanism for sending data over the

USB to your Serial Monitor. Go to the book's code repository and check the related sketch code.

As you will notice in the code, a SoftwareSerial object (mySerial) is initialized by defining the digital pins 2 and 3 as Rx and Tx pins respectively. These pins will be used for the communication with the RFID Reader and thus you need to reconfigure your circuit appropriately. The latter also means that you have to replace any communication with the Reader using the Serial.read() method in the previous sketch with mySerial.read() as well.

Using the I2C Protocol

The I2C (Inter-Integrated Circuit) is another standard designed for communication between embedded devices and sensors. It can be also used to enable communicate between several Arduino boards (see next Section for more details). Some of the special characteristics of the I2C protocol are that it requires two ports-pins for communication, the SCL (clock line used for synchronization) and the SDA (the data line) and that devices are connected in a master-slave mode. The master device coordinates the transfer of information on the slave devices. There always has to be a master device and when it comes to sensor communication, the Arduino is usually the master one. In order the master device to recognize and coordinate the slaves, the latter must have a unique address number. All commercial sensor devices that use the I2C come with their own address that you need to retrieve it from the vendor's specification and use it in your code.

In case you use I2C to communicate various Arduino boards with each other, you need to set addresses yourself to the slave nodes. On a standard Arduino board like the Uno, the analog 5 is used for SCL and analog pin 4 for SDL.

A main disadvantage of the I2C protocol is that the data rate is slower than other protocols (e.g., the SPI discussed at the end of the Chapter) and that data can only travel in one direction at a time.

When connecting an I2C sensor/device, also keep in mind that it is necessary to connect pull-up resistors to the connections to ensure reliable transmission of signals (unless your sensor board integrates it already).

Communicating with a Digital Pressure Sensor

Let's see an example of I2C communication using a digital pressure sensor. The latter is also known as a barometer and can provide you with readings about the temperature and pressure in the atmosphere that can give indications about the weather condition. A low pressure typically means cloudier, rainier, and colder weather, whereas high pressure generally means clearer and sunnier weather.

An easy to use digital pressure sensor based on the I2C protocol is the BMP085 module from Sparkfun (see Figure 5-8). It comes assembled in a breakout board that includes all the essential components for the sensor to operate (like a pull-up resistor) and can be directly connected to your Arduino board.

To build the circuit just connect the Gnd and 3.3V pins (last two) and the SCL and SDA pins to your Arduino board as in Figure 5-9. SCL goes to pin A5 and SDA to pin A4.

Talking I2C through The Code

First thing you need to know about programming an Arduino sketch that communicates over the I2C protocol, is that you have to study the device's datasheet. There is no way you can connect pins together, read some bytes and be able to turn them directly into useful information unless you can figure out the structure of the messages and how the device sends them.

Reading through the vendor's datasheet will reveal you information about the operating voltage, circuit setup, the I2C address of the device, operational principles (such as essential calibration for proper value reading, time delays that need to be introduced before reading from the device, etc.) and most importantly how to trigger the device to sense and how to ask for readings from it. The latter information is usually revealed in the datasheet in sections talking about the timing diagram and control register values in I2C-related sections.

Figure 5-8. A barometric pressure sensor by Sparkfun (image courtesy of Sparkfun)

Regarding the BMP085 sensor (manufactured by Bosch) the datasheet (you can look for it under the product description on the Sparkfun product page) describes the process of reading the temperature and pressure as twofold: Initially you have to read the internally stored (in EEPROM memory) calibration values (eleven 16-bit calibration coefficients) and then move on reading the data. The section in the datasheet that describes the data reading is the following:

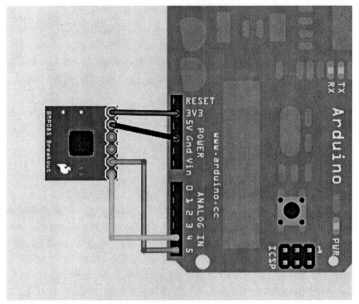

Figure 5-9. A barometric pressure sensor by Sparkfun (image courtesy of Sparfun)

To read out the temperature data word UT (16 bit), the pressure data word UP (16 to 19 bit) and the EEPROM data proceed as follows:
After the start condition the master sends the module address write command and register address. The register address selects the read register:
EEPROM data registers   0xAA to 0xBF
Temperature of pressure value UT or UP   0xF6 (MSB), 0xF7 (LSB), optionally 0xF8 (XLSB)

Then the master sends a restart condition followed by the module address read that will be acknowledged by the BMP085 (ACKS). The BMP085 sends first the 8 MSB, acknowledged by the master (ACKM), then the 8 LSB. The master sends a « not acknowledge » (NACKM) and finally a stop condition.

So according to the datasheet you will need to read 8-bit (1 byte) values (the MSB and LSB) and 16-bit (2 byte) values (the temperature and pressure readings) from the BMP085 module. The following listing presents two functions that do so. The code is adopted from the sketch provided by Jim Lindblom of Sparkfun that demonstrates the communication of Arduino with the BMp085 module. You can find the complete code in the book's code repository (**http://www.buildinginternetofthings.com**).

As indicated in the code, Arduino communicates with the sensor through its I2C address. The latter has to be defined in the beginning of your code as:

```
#define BMP085_ADDRESS 0x77  // I2C address of BMP085
```

Each function has the address to be read as an input parameter (the address can be an EEPROM data register and/or 0xF6 or 0xF7 for asking the temperature or the pressure respectively, as the datasheet describes). It initiates the communication using the Wire library with the specified I2C module address and it transmits the requested address.

It then requests tor receive the respective data size of data (1 or 2 bytes) and returns the incoming byte from the module. In the case of bmp085ReadInt() it receives two separate bytes from module (as MSB and LSB) and then returns them as one 16-bit value.

Since you have implemented the essential functions for reading and sending byte values to the module, you can implement the essential functions for calibrating the module and reading the temperature and pressure values.

The first function (bmp085Calibration()) reads the calibration values form the device's EEPROM and stores them into program variables. It uses the bmp085ReadInt() function by passing the EEPROM data register addresses (from 0xAA up to 0xBE) as input parameters.

The last two functions read the temperature and pressure values. As with the previous functions, they initialize the communication with the module defining its I2C address. They send the appropriate commands and wait the essential amount of time as defined by the datasheet. Then they send the byte values that correspond to the temperature and pressure reading using the bmp085ReadInt() function.

The complete code for reading the BMP065 module temperature and pressure values can also be found here: **http://www.sparkfun.com/tutorial/Barometric/BMP085_Example_ Code.pde**

Read Data From Another Arduino Board

Why would you need to use a second Arduino board to read data from it? One answer is that you might find yourself in a situation where 6 analog input pins are not enough (of course depending on your project you might run out of digital pins as well) and you can use the second board as an extension. Another idea is that you might use the first board for some other task and you might need a second dedicated board for sensing only (like in reprogramming your Arduino through the Cloud as described in Chapter 10).

Let's see how you can make this.

You will need two Arduino boards and three jumper wires. The two of them will be used to connect the communication pins between the two boards and the third one to make a common Ground connection between them.

One Arduino board will be the master device and the other the slave one. Depending on your needs, you can set as slave the board that reads some kind of sensor and forwards the readings to the master one. For this example you will use the LDR sensor as in the beginning of the Chapter in order to detect light changes. The sensor will be attached to the slave board and the readings will be forwarded to the master board.

Apart from the circuit setup (where you will connect the sensor on the slave Arduino), you also need to run a master and a slave sketch to each board respectively.

The Circuit Setup

Connect the two boards as in Figure 5-10: Analog port 4 of the master Arduino is connected to the analog port 4 and respectively analog port 5 of the master Arduino is connected to port 5 of the slave one. Make sure you connect the Ground ports of the two boards together as well. On the slave connect a light sensor (LDR) on port 0 with a 10K Ohm resistor as illustrated.

One good idea is to program the slave Arduino board first before making any connections for your convenience.

The Master Arduino Code

The sketch for the master Arduino board requests information from the slave board and prints it over the Serial Monitor.

Listing 5-4. The Master Arduino Sketch

```
#include <Wire.h>
void setup() {
  // Initialize the I2C bus
  Wire.begin();
  Serial.begin(9600);  // start serial for output
}

void loop() {
  // request 1 byte from slave arduino with address #2
  Wire.requestFrom(2, 1);
  // while there are data arriving from slave Arduino
  while(Wire.available()) {
```

```
    // receive a byte as character and print it
    char c = Wire.read();
    Serial.print(c);
  }

  delay(1000);
}
```

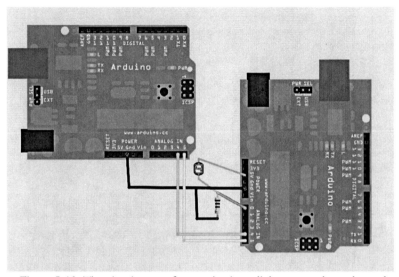

Figure 5-10. The circuit setup for monitoring a light sensor through another (master) Arduino.

It initializes the I2C protocol using the Wire library (already imported in your Arduino IDE) and within the loop() function it requests data from the slave Arduino. To do so, it defined the address of the slave board ('2' in this case) and requests to read 1 byte at a time (since the slave will be sending sensor readings in 1 byte variable length). Then it checks for incoming data, using the Wire.available() method, and reads them through Wire.read().

The Slave Arduino Code
The slave sketch needs to implement a couple of things more. It has to monitor for requests from the server, read the light sensor input from the analog port and forward the latter to the server. Check the following listing for the complete slave code and its explanation at the end.

Listing 5-5. The Slave Arduino Code

```
#include <Wire.h>

//Store sensor readings in this variable
int sensorValue;
```

```
void setup(){
  // Initialize the I2C bus and assign address number #2 to
the Slave Arduino
  Wire.begin(2);
  // register event listener function
  Wire.onRequest(requestEvent);
}

void loop(){
  //Read the LDR sensor in pin 0
  sensorValue = analogRead(A0);
  delay(100);
}

// This is the event listerner function
//that is executed whenever data is requested by Master
Arduino
void requestEvent(){
  //Send the sensor reading to the Master Arduino (1 byte
value)
  Wire.write(sensorValue);
}
```

At first the slave has to define its address in order to be identified by the master. This is done in the setup() function by using the Wire.begin() function. You probably notice that the begin() method is pretty common in the Arduino world for initializing communication protocols. The defined address for the slave board is '2', as also used in the master sketch.

Then you need to define the slave to send sensor readings when there is a request from the master Arduino. This is done using an event listener function. You use the Wire.onRequest() method that takes a sketch-defined function as an input. The function will be automatically invoked (through the Wire library) whenever there is a new request from the master.

You define and implement the function (named requestEvent()) at the end of the sketch. It just prints using the Wire protocol the sensor reading stored in a variable (sensorValue).

The loop() function reads the LDR value the same way it does in the similar analog sensor example.

The SPI Protocol

The Serial Peripheral Interface (SPI) protocol is the most standard protocol in inter-device communication. Its main features (and advantages compared to I2C) are that it runs at a higher data rate, and it has separate input and output connections, so it can send and receive at the same time (this feature is called full duplex mode). It uses one additional line per device to select the active device, so more connections are required if you have many devices to connect. Most Arduino projects use SPI devices for high data rate applications such as Ethernet and WiFi modules (I2C is more typically used with sensors that don't need to send a lot of data).

Like the I2C, it operates in a master/slave mode. So you need at least one master device and one or more slave ones.

SPI has separate input (labeled "MOSI" from Master Output, Slave Input) and output (labeled "MISO") lines and a clock line (labeled "SCLK" from Serial Clock Line). These three lines are connected to the respective lines on one or more slaves. Slaves are identified by signaling with the Slave Select (SS) line.

On an Arduino Uno board digital pin 13 is used for Clock (SCLK), pin 12 for data out (MISO), pin 11 for data in (MOSI) and pin 10 for the Slave Select (SS) line.

In order the master device to begin a communication, it first configures the clock, using a frequency less than or equal to the maximum frequency the slave devices support. During each clock cycle, a full duplex data transmission occurs between the master and the slave: the master sends a bit on the MOSI line and the slave reads it from that same line. Then the slave sends a bit on the MISO line and the master reads it from that same line as well.

You may find the SPI protocol communication a bit complex; it is, but luckily most of the device (communication or sensor modules) vendors will provide you with code samples that demonstrate the implementation of this master/slave interaction.

Summary

Enabling your Arduino to sense the world is the first big step for building IoT projects. Chapter 2 has given you many ideas of what kind of sensors you can connect to your Arduino board. This Chapter has presented you how to connect various sensors on your Arduino board, digital and analog ones. You have also been taught about the different communication protocols that vendors use for digital sensors (Serial, I2C, SPI) and you have seen examples of each case. The I2C protocol implementation provided by Arduino allows you also to connect two or more boards together and transfer sensor reading from one to another.

This is the basic knowledge you need to sense your environment using your favorite Arduino board. What comes next? Learn how to communicate with your Android phone and the Internet.

6 TALKING TO YOUR ANDROID PHONE WITH THE ARDUINO

There are several reasons you might want to connect your Arduino with your Android phone. You may want to control your Arduino from your phone, as a kind or remote control. Another case is that you can use your phone to store sensor data from your Arduino. Or simply, you might use your phone to forward data to the Internet. The latter is very important since it can prove a nice solution when building IoT networks and applications. The phone can be a gateway to receive and forward data to the Cloud when direct wired or wireless interfaces are not available for your Arduino.

This chapter will teach you how to communicate your Arduino with your Android phone. Based on the interfaces a phone might have, you will explore both the communication using the USB port and connection using the Bluetooth technology. The recently introduced Android Open Accessory Development Kit (ADK) by Google can be a candidate technology for the first option. As an alternative you will explore more the Android Debug Bridge (ADB) and the supporting hardware. The second way is based on the default Bluetooth interface that exists in almost every Android phone and serial communication with the Arduino. The chapter also demonstrates how to use these technologies in order to Tweet sensor data through your phone, control an Arduino port using SMS messages and store air quality data on your phone.

Connecting Arduino with a mobile device

There have been several (successful) attempts in the past to connect an Arduino board with a mobile device. People wanted always a kind of remote control for their Arduino project and were seeking something more useful than using infrared or RF modules. Especially at the point Android was introduced and mobile device development got much easier than it was.

Before the introduction of the ADK by Google, only a few options were available for connecting the two devices. One could use either Bluetooth and 'talk' over the air using serial communication protocol, or alternatively 'hack' the USB interface of Android. Such an example is a project by Cellbots (www.cellbots.com) hosted by instructubles.com that requires a rooted Android phone that uses serial output to control an Arduino robot (see Figure 6-1). A totally different approach was followed by Jeffrey Nelson who had the idea to control his Arduino through the DMTF dial tones of the phone (see Figure 6-2). He can send commands from a PC to the phone over WiFi or 3G access, and the phone generates corresponding dial tones. The Arduino interprets the tones and servomotors are driven this way.

Regarding Bluetooth communication, initially you had to implement your own serial-based communication protocol between the phone and the Android until the Amarino toolkit (http://www.amarino-toolkit.net/) was introduced. It is still the best way to implement Bluetooth communication without spending much effort and time in coding the serial protocol. Still however, connection between phone and Arduino through Bluetooth is not very straightforward and can require additional steps like setting up properly a Bluetooth serial modem and pairing with the phone. Also, Bluetooth consumes much more energy than a standard USB connection, so this is another issue to consider when selecting which way you will go in your projects.

Figure 6-1. A serial-based communication of an Arduino clone and a 1st generation Android phone (image courtesy of instructables.com).

Figure 6-2. The 'Forknife': A robot controlled remotely with the help of an Android phone. One of the first attempts to communicate Android with Arduino. This one uses DMTF dial tones generated by an Android app as means to control outputs from the Arduino.

After the introduction of the Google ADK, things got much easier and direct in connecting an Android phone with an Arduino-enabled board. Still special hardware is needed, meaning that there is no way to directly interface a plane Arduino board (like the Uno) with an Android device. You need an ADK compatible development board.

According to Google, the following list of board vendors are currently producing Android Open Accessory compatible development boards:

- The official Arduino team has released the Arduino Mega ADK that is based on the ATmega2560 and supports the ADK firmware.
- DIY Drones provides an Arduino-compatible board geared towards RC (radio controlled) and UAV (unmanned aerial vehicle) enthusiasts.
- Microchip provides a PIC based USB microcontroller board.
- Modern Device provides an Arduino-compatible board that supports the ADK firmware.
- RT Corp provides an Arduino-compatible board based on the Android ADK board design.
- SeeedStudio provides an Arduino-compatible board that supports the ADK firmware (see Figure 6-3).
- SparkFun's IOIO board now has beta support for the ADK firmware.

In addition to the boards (that offer complete Arduino features and functionality) there is the Circuits@Home USB Host Shield that can be used with any current Arduino board and offer ADK functionality.

The list with supported boards is still expanding and probably by the time you read this chapter more modules have already been added.

Once again, there are several benefits of interfacing Arduino with an Android phone. Here is a list with some of them:

- The phone can be an excellent remote control for the Arduino: You can control the outputs of the Arduino board by using the touch screen of the phone and/or the acceleration and tilt sensors.
- You can provide Internet connectivity to the Arduino board: Since you can forward data from the board to the phone you can also program the phone to save the data online. Quite useful feature especially when managing sensor data on the Cloud. You can also use the phone to receive data from the Internet and forward them to the board.
- You can store and process sensor data on the phone: Phones generally can save lots of data either in internal memory or through micro SD cards. This is quite useful since you can avoid the cost and the effort to save and read data on an SD card through the

Arduino. In addition, you might want to do some processing with the data, like try to predict things by applying data classification techniques (like neural networks, etc.) Arduino has by no means the processing power to perform anything like that, whereas the phone is much more capable.

- You can utilize the additional hardware modules of the phone: For example you can use the GPS sensor and add location information to your sensor data.

The projects of this book will give you all the knowledge you need to make your own projects with your favorite Arduino board and an Android phone.

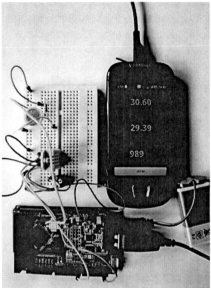

Figure 6-3. An Android phone connected to a Seeeduino Main ADK board (Part of a project you will explore in Chapter 8).

The Android mobile OS

The Android OS has been introduced by Google as an operating system for mobile phones, Tablets and even Notebooks and TV sets. The first Android-enabled phone introduced to the market was the HTC Dream in 2008. Since then, there are dozens of models and various vendors (Samsung, HTC, LG, etc.) that feature Android OS.

One of the major benefits and breakthroughs of Android is the great access to phone resources the developers have. You can combine information from the web with data on mobile phone (like contacts, calendar, or geographic location) and build an application that can communicate over the Network with Cloud applications, send and receive SMSs and even make calls.

Some of the many advantages to developing applications for the Android platform are listed here:

- The development tools for the platform are free to download and use.
- The Android OS is an open-source platform based on the Linux kernel and multiple open-source libraries.
- Very few restrictions are placed on the content or functionality allowed in Google's Android Market, and developers are free to distribute their applications through other distribution channels as well.
- There are a wide variety of hardware devices powered by the Android OS, including many different phones, tablet computers and (very soon) other devices. Development for the platform can be made on any OS (Windows, Mac OS or Linux).
- There is a great number of users with Android-based phones.
- The best one: You can easily interface it with Arduino! It is not that easy (yet) to achieve the latter with other mobile platforms.

To write your own Android mobile phone applications, you'll first need to collect the required tools and set up an appropriate development environment on your PC or Mac. No matter which operating system you are using, you will need essentially the same set of tools:

- The Eclipse Integrated Development Environment
- Sun's Java Development Kit (JDK)
- The Android Software Developer's Kit (SDK)
- A special Eclipse plug in: the Android Developer Tool (ADT)

Demonstrating how to set up a development environment for Android and how to develop mobile applications is out of the context of this book. If you are still in the need for guidance on Android development you can consult many of the books for available for Android programming.

Let's move on to interfacing Arduino with an Android phone!

Communicating using Bluetooth

As discussed in previous section, there two ways to communicate your phone with your Arduino board and allow full data exchange; either using Bluetooth or a USB connection. Bluetooth was for some time the only option, since the USB communication required some additional skills and/or rooting your mobile.

Hardware

In order to communicate with an Android phone you obviously need to add Bluetooth functionality to your Arduino board. There are several ways to do that: you can connect a Bluetooth serial (modem) module directly to your Arduino, you can use a shield and you can also use the official Arduino Bluetooth board provided by the Arduino team.

For the first option you can easily find many modules to use. Quite popular for Arduino projects is the BlueSMiRF Silver modem (see Figure 6-4) and the Bluetooth Bee by Seeedstudio (see Figure 6-5). The first can be connected directly to your Arduino Rx/Tx pins and the second can be placed over a shield.

Figure 6-4. The BlueSMiRF Bluetooth serial modem. It can be directly connected to your Arduino Rx/Tx pins (image courtesy of Sparkfun).

Figure 6-5. The Bluetooth Bee by Seeedstudio. It can be placed easily on an appropriate shield or over an Arduino Fio board (image courtesy of Seeedstudio).

You can select any type of module and vendor based on price, size and 'pluggable' (how easy you can connect it with your Arduino) requirements of your project.

Configure your Bluetooth serial modem

Depending on the Bluetooth modem vendor you might need to configure properly your module before your can use it in your projects. To do so you can write a small sketch that will send the appropriate commands to the module over Serial interface and then connect the module with your Arduino.

For example, the BlueSMiRF module needs the following commands for proper initialization:

```
void setup()
```

```
{
  // use the baud rate your bluetooth module is configured
to
  Serial.begin(9600);
  Serial.write("AT+BTLNM=\"ArduinoAndTheCloud\"");
  delay(1000);
  Serial.write("AT+BTAUT=1, 1");
  delay(1000);
  Serial.write("AT+BFLS");
}
```

This code will initialize the modem with the name "ArduinoAndTheCloud", it will enable Automatic Connection Mode and also will save these settings permanently to Flash memory.

Likewise, the following code initializes the Bluetooth Bee module and sets it in client (slave) mode, it sets the device name, it enables auto-connect, it enables paired-devices (your phone) to contact it, it sets the pair pin to 0000 and finally it puts the module to inquiry mode (so that your phone can find it).

```
void setup()
{
  // use the baud rate your bluetooth module is configured
to
  Serial.begin(38400);
  delay(1000);
  Serial.write("\r\n+STWMOD=0\r\n");
  Serial.write("\r\n+STNA=ArduinoAndTheCloud\r\n");
  Serial.write("\r\n+STAUTO=0\r\n");
  Serial.write("\r\n+STOAUT=1\r\n");
  Serial.write("\r\n +STPIN=0000\r\n");
  delay(2000); // This delay is required.
  Serial.write("\r\n+INQ=1\r\n");
  delay(2000); // This delay is required.
}
```

Keep in mind that you need to pay attention to the default baud rate of your Bluetooth module, as set by the vendor's instruction, otherwise the initialization will fail and most likely you will not be able to pair the module with your phone and communicate with it at all!

Software

The hardware we have been discussing about consists of Bluetooth serial modems. This suggests that the communication can be achieved using the default Arduino serial library (sending data through Serial.write() command in the simplest way) or any other alternative library (like the NewSoftSerial library). Considering that you have successfully initialized your Bluetooth module and paired it with your phone you need to do the following:

- Write a sketch that will implement all the serial communication using your own protocol.
- Write an android app that will implement all the essential code for receiving and sending fata over Bluetooth.

There is a software library that abstracts the latter communication, the Amarino-toolkit (http://www.amarino-toolkit.net). The toolkit enables users to connect Android-driven mobile devices with Arduino microcontrollers via Bluetooth. It provides easy access to internal phone events, which can be further, processed on the Arduino and has great features like:

- Allowing users to control multiple Bluetooth devices in parallel
- There are default Android apps users can use directly and visual feedback from sensors and for events sent from the phone
- Providing a plug-in concept for users to integrate their own events with Amarino (and using its visualizer for feedback)
- The library supports for Android 1.6 and 2.x devices
- It includes an Android library ('MeetAndroid') to talk to Amarino from custom apps.

You only need to install the Arduino library and use it in your sketches. Direct methods are available for communicating with the phone. Same applies for your Android app. Using the appropriate library you get methods that listen for incoming data and send data to your phone using very simple code.

To sum up, in order your Android phone to communicate with your Arduino board you need the following:

1. A Bluetooth serial module (or a Bluetooth Arduino) for your Arduino board
2. You need to configure your Bluetooth module appropriately so it can be paired with your phone
3. You need to install the Amarino library for the Arduino ('MeetAndroid') to your Arduino environment
4. You need to write a sketch that will use the MeetAndroid library's methods to send and receive data (instructions on how to do this are given in the following example projects)
5. You need to write an Android app that will use the Amarino library to receive and send data from and to the Arduino via Bluetooth.

Let's move on with your first project and see how you can achieve this communication over the air and how you can make something useful out of it!

Project 1- Send Air Quality data to your phone

How about being able to monitor the quality of the air you breath and you store the data on your phone, so that you can latter forward them to the Internet. For this project you will use a Carbon Monoxide (CO) sensor, suitable for sensing CO concentrations in the air. A suitable sensor is the MQ-7 CO sensor (available also from Sparkfun electronics) that can detect CO concentrations anywhere from 20 to 2000ppm. Usually such sensors are analog sensors and are connected to the Arduino directly through an appropriate dedicated analog pin. However, due to their operational need, several pins need to be connected to a 5V power source and the ground, so using a Gas Sensor Breakout Board from Sparkfun will help you avoid some extra pin soldering.

The Circuit

The basic parts for this project are:

- A Bluetooth serial modem (in case you don't have an Arduino Bluetooth board)
- An Arduino board
- An Air quality – Gas sensor like the MQ-7 CO sensor
- Some wires to connect everything together
- (Optionally) The Gas Sensor Breakout Board from Sparkfun that will assist you with the connection of the sensor.
- Your Android phone

You will need to connect all components together based on the circuit setup illustrated in Figure 6-6. Make sure you connect the 3.3V pin of Arduino to your modem's Vcc port and the Rx port of the modem with the Tx pin of the Arduino and the Tx port of the modem with the Rx pin of the Arduino respectively. This makes sense since the modem needs to send data (Tx) from where the Arduino needs to read data (Rx) and vice versa.

Also make sure that the Gas sensor B1 (power pin) is connected to the 5V pin of the Arduino. Once you got the circuit set up right you can move on with the Arduino sketch for reading sensor values and sending them to your phone over Bluetooth.

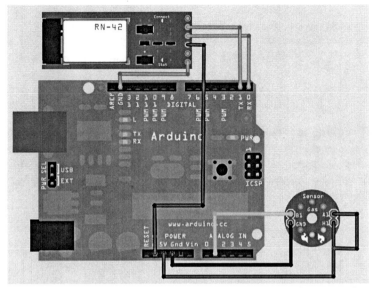

Figure 6-6. The circuit setup for connecting the Bluetooth modem and the gas sensor to the Arduino. Make sure you power the modem with 3.3V from the Arduino board and you connect the Rx pin of the modem with the Tx port of the Arduino, and the Tx pin of the modem with the Rx port of the Arduino respectively.

The Arduino Code

Make sure you have properly installed the 'MeetAndroid' library by Amarino toolkit in your Arduino environment. Then just upload the following sketch to your board.

Listing 6-1. This is the Arduino Code for reading Gas sensor values and forwarding them to the Android phone via Bluetooth

```
#include <MeetAndroid.h>

MeetAndroid meetAndroid;
int GASsensor = 1;

void setup()
{
  // use the baud rate your bluetooth module is configured
to
  Serial.begin(57600);

  // we initialize the analog sensor pin as an input pin
  pinMode(GASsensor , INPUT);
}

void loop()
{
```

```
  // read input pin and send result to Android
  meetAndroid.send(analogRead(GASsensor ));
  // Send a reading every 10 minutes
  delay(600000);
}
```

By running the sketch on the Arduino you will probably notice some of the LEDs of the Bluetooth modem to go on and off or being constantly on. This has to do with the operational mode of your modem, indicating that it is in inquiry mode, it has paired with the phone and/or it is sending/receiving data.

The Arduino Code – Code Review
At the beginning of your sketch you need to import the installed MeetAndroid library:

```
#include <MeetAndroid.h>
```

Then you create the MeetAndroid object that will provide us with the necessary methods to communicate with the phone over Bluetooth.

```
MeetAndroid meetAndroid;
```

You also set analog pin 1 of Arduino as connected pin to sensor's analog output.

```
int GASsensor = 1;
```

You need initialize the Serial communication to the specific baudrate of your Bluetooth module. Notice that the Serial library is used (as default) to communicate with the Bluetooth modem.

```
void setup()
{
  // use the baud rate your bluetooth module is configured
to
  Serial.begin(57600);
  You also initialize the analog sensor pin as an input pin
  pinMode(sensor, INPUT);
```

Finally, the loop() method contains all you need to send data to the phone. Amarino through the MeetAndroid library simplifies very much the communication by just invoking the send() method and passing the data you want to send as an argument:

```
void loop()
{

  // read input pin and send result to Android
```

```
meetAndroid.send(analogRead(sensor));
/// Send a reading every 10 minutes
delay(600000);
}
```

Install Amarino on your phone

Before moving on to the Android app that will receive data from the board, it is necessary to pair/connect your Bluetooth modem with your phone. The most easy and convenient way is to do it directly from your Android phone. Once there, you need to install the Amarino Android application. The latter is also essential for your own Android apps to use the Amarino library and work. You can find the latest Amarino – Android application version in the Download section of the Amarino toolkit web site (http://www.amarino-toolkit.net/).

Once installed, the very first step you have to do is to search for your paired Bluetooth modem. To do that hit the "Add BT Device" button and wait until your Arduino Bluetooth module pops up. Then just click the 'Connect' button. When the green led next to your modem's name goes on, it means the application has successfully found your modem and is connected to it. The led indicators on the modem will also light accordingly. From this point you can move on to creating your own Amarino-based Android application that will be able to communicate with your Arduino board over Bluetooth!

The Android Code

You need to create a new project in Eclipse environment and add the Amarino library (AmarinoLibrary.jar) to your project's path. Then create a new activity and use the android code found in the code repository for this Chapter.

You should not forget to add the 'Internet' permission to the AndroidManifest.xml of your project. You can check at the book's code repository (**http://www.buildinginternetofthings.com**) the full source code of an Android project that receives data through Bluetooth and saves them into a text file in the SD card memory of the phone.

The Android Code – Code Review

In the beginning of the code you will notice the import of the appropriate Android libraries and the Amarino library as well.

```
import android.app.Activity;
import android.content.BroadcastReceiver;
import android.content.Context;
import android.content.Intent;
import android.content.IntentFilter;
import android.os.Bundle;
```

```
import at.abraxas.amarino.Amarino;
import at.abraxas.amarino.AmarinoIntent;
```

Next you start with the main Activity class

```
public class ArduinoGasSensor extends Activity
```

You set a variable for the MAC address of your Bluetooth module. You need to change this according to your Bluetooth device address.

```
    private static final String DEVICE_ADDRESS =
"00:18:E4:0C:68:C5";
```

The following line of code instantiates an ArduinoReceiver object based on the class you implement later on the code. This is the class that will run as a background service and listen for incoming requests from the Arduino.

```
    private ArduinoReceiver arduinoReceiver = new
ArduinoReceiver();
```

The usual for Android activities onCreate() method does not contain anything Amarino-specific.

```
    /** Called when the activity is first created. */
    @Override
    public void onCreate(Bundle savedInstanceState) {
        super.onCreate(savedInstanceState);
        setContentView(R.layout.main);
    }
```

Instead, you do implement the onStart() and onStop(0 methods since they are quite essential for the operation of the Amarino. The onStart() invoked when the Activity starts will register the arduinoReceiver object you have implemented to the Android system so that it can be activated and listen for events.   You also connect the Amarino with the Bluetooth modem through the specified MAC Address.

```
    @Override
        protected void onStart() {
            super.onStart();
            // in order to receive broadcasted intents you need
            //to register the receiver
            registerReceiver(arduinoReceiver, new
            IntentFilter(AmarinoIntent.ACTION_RECEIVED));

            // this is how you tell Amarino to connect to a
            //specific BT device from within your own code
            Amarino.connect(this, DEVICE_ADDRESS);
        }
```

The onStop() method (invoked when application is exited) will respectively disconnect the device and unregister the arduinoReceiver.

```
@Override
protected void onStop() {
    super.onStop();

    // if you connect in onStart() you must not forget
    //to disconnect when your app is closed
    Amarino.disconnect(this, DEVICE_ADDRESS);

    // do never forget to unregister a registered
    //receiver
    unregisterReceiver(arduinoReceiver);
}
```

The last part of the code is the most important one. It is contains the implementation of the arduinoReciever object, the ArduinoReceiver class. This class is responsible for catching broadcasted Amarino events.

```
public class ArduinoReceiver extends BroadcastReceiver
```

Since the class extends the BroacastReceiver Android class (which allows it to listen for events), you need to implement the onReceive method.

```
@Override
    public void onReceive(Context context, Intent
    intent) {
        String data = null;
```

Here you handle all incoming data from the Amarino. You define the type of data that is added to the intent.

```
final int dataType =
intent.getIntExtra(AmarinoIntent.EXTRA_DATA_TYPE, -1);
```

You only expect String data though, but it is better to check if really string was sent.

```
if (dataType == AmarinoIntent.STRING_EXTRA)
```

and you retrieve the String data from the Amarino Intent.

```
data = intent.getStringExtra(AmarinoIntent.EXTRA_DATA);
```

You make a check in case data is null and then you just do anything with it:

```
if (data != null){
    //Do something with the data Arduino has sent you!
}
```

You can extend the Android code to add location information. This way you can monitor the air quality of places you visit!

Communicating using USB

As we discussed in the beginning of this chapter, the USB communication is the alternative to the Bluetooth for making your Android phone talk to your Arduino board! Currently, there are two ways to do this that both depend on different hardware and software.

Using the Google ADK

As we already have discussed, the Google ADK is supported by several board vendors. By the time it was introduced, one of the first compatible boards released was the official Arduino Mega ADK (see Figure 6-7). The Arduino Mega ADK is a microcontroller board based on the ATmega2560. It has a USB host interface to connect with Android based phones and it is full compatible with Android's ADK examples posted by Google. It also features 54 digital input/output pins (14 of them can be used as PWM outputs), 16 analog inputs, and 4 UART ports. Similarly to other board it also includes a 16 MHz crystal oscillator, a USB connection, a power jack, an ICSP header, and a reset button.

In order to use an ADK compatible board you need to download the following items and set up in your development environment:

- CapSense library: contains the libraries to sense human capacitance. This is needed for the capacitive button that is located on the ADK shield.
- The ADK package: contains the firmware for the ADK board and hardware design files for the ADK board and shield.

More information on where to retrieve the latter and setup instructions can be obtained from here: http://developer.android.com/guide/topics/usb/adk.html

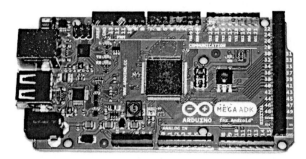

Figure 6-7. The official Arduino Mega ADK board.

In case you cannot find the Mega ADK Board in your Arduino IDE Board selection, you might need to change the 'boards.txt' file in your Arduino directory. More instructions on this here: http://www.arduino.cc/en/Main/ArduinoBoardADK

A drawback of the ADK (compared to the ADB method described in next section) is that it requires phones running Android 2.3.4 and later. Since ADB is more interoperable (supporting even phones running Android 2.5) most of the projects presented will be based on the ADB.

Using the ADB

According to Google, The Android Debug Bridge (ADB) has been originally provided to the developers of Android apps as a (command line) tool for communicating with emulated instances or real computer-connected Android devices. Fortunately, you can use ADB to connect your Arduino board and your Android phone and use the ADB over a USB connection.

ADB provides both the tools that are executed on your computer and those running on your phone or on the emulator so that the latter can communicate.

There are special ADB-based implementations that provide us with the appropriate libraries for making such connections. Such an implementation that you will be using in the projects of this books is MicroBridge.

MicroBridge allows any kind of Android device to talk directly to USB host enabled microcontroller units, thereby enabling phones to actuate servos, drive DC motors, talk to I2C and SPI devices, read analog and digital sensors, and so forth. A great benefit of using MicroBridge (against ADK) is the fact that it works on Android 1.5 and upwards. This means you can use even a first generation Android phone with your Arduino board!

The ADB protocol supports opening a shell on the phone or issuing shell commands directly, pushing files to the phone, reading debugging logs (logcat), and forwarding TCP ports and Unix sockets. Using TCP sockets it's possible to establish bidirectional communication links between an Arduino and an Android device. The Android application listens on a port, and the Arduino connects to that port over ADB. When the connection is lost (e.g., in case the application has crashed or because the USB device is unplugged) MicroBridge periodically tries to reconnect. When the device is plugged back in, the link is automatically reestablished. MicroBridge enables you to do pretty much anything you can when using the ADB tool. There are two versions available, one written in C, and a C++/Wiring port for Arduino.

The basic idea in an Arduino sketch that uses ADB is that you make a TCP connection to a specific port (where actually the ADB Android server listens to) and you use an event handler. In the main loop you 'poll' the connection for incoming data. You handle the data in the implementation function of the event handler. More details on how this actually works will be given in the following project examples.

Hardware

MicroBridge is compatible with various Arduino-based boards or extension shields. The most popular are the Seeeduino ADK Main Board (that is also used in the presented projects), the USB Host Shield by Oleg Mazurov, and of course the original Arduino ADK Board.

The Seeeduino ADK Main Board (see Figure 6-8) supports Android devices v1.5 using MicroBridge and v2.3.4 and above with Google ADK. The main benefits of this board is that it works like Arduino Mega 2560 with inbuilt USB Host Shield and features both ADK and ADB communication with Android phones.

The USB Host Shield (see Figure 6-9) is a much more economic solution, since it is a shield only. It is compatible however with any Arduino basic and Mega board.

In order to use the MicroBridge ADB library you need to download the latest version and include it into the libraries folder of your Arduino installation. You can also download a version of the library from the book's online code repository.

In case you face compilation issues with the MicroBridge ADB and Arduino version 1.x you can use the version included in the online code repository.

Figure 6-8. The Seeeduino ADK Main Board. Full Mega2560 compatible Arduino board that supports both the Google ADK and the MicroBridge ADB. The phone is connected through the main USB port, whereas the mini usb is used for programming the board (image courtesy of Seeedstudio).

Figure 6-9. The USB Host Shield by Circuits@Home. It is a shield only solution for ADB communication and can be used in any Arduino basic and Mega board.

Project 2 - Tweet Sensor readings through your phone

Let's make something more advanced, which also involves communication with the Internet through your phone. How about letting your friends know the temperature of your room via Twitter ? The idea is that your Arduino will read the temperature sensor and send the reading to your phone, which then will post a Twitter status update on your account.

Twitter allows third applications to communicate with it (and read/post messages) only when the latter are registered to it. Therefore, for this project (apart from a Twitter account) you also need to create a Twitter application. To do so, go to Twitter Apps page (http://www.twitter.com/apps) and register your application (Use: Create a New Application). Make sure you pick a unique name for your application!

Add a description of your project (e.g., Tweet my Temp) and add a dummy URL for the Website (e.g., http://www.mysite.com), like in Figure 6-10. After you create the Twitter application you will get to the next screen where you can set some settings about your application and retrieve important information like the access keys. Once there, go to 'Settings' tab and select 'Read and Write' for the Application Type (so that your Android app can post messages). Update the new settings and go back to the 'Details' page and click the 'Create my access token'. This step is the last essential one so that your app can post updates to Twitter. From all the information contained in the 'Details' page you will need the 'Consumer key', the Consumer secret, the Access token and the Access token secret. Note them down and you can move on with the circuit setup and the project code!

Figure 6-10. Setting up the Twitter application for enabling your Android app to send Tweets to your account.

Figure 6-11. Important information (keys and tokens) that you need to use in your Android code.

The Circuit

You will need a temperature analog sensor (like 'Analog Devices TMP36'), a 10K Ohm resistor, an Arduino ADB board, a USB cable and an Android phone. Optionally you can use a Breadboard for connecting everything together more conveniently.

Connect the resistor and the sensor to the Arduino board as illustrated in the following figure. The Arduino itself will power the sensor, so you only need a power source the Arduino (not illustrated in the figure).

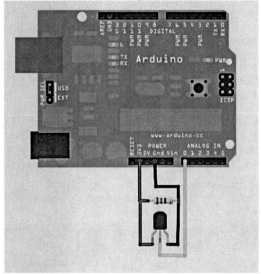

Figure 6-12. Illustration of the circuit setup for connecting the temperature sensor to the Arduino

The Arduino Code

Listing 6-2. This Is the Arduino Code for receiving Events from the Android phone

```
#include <SPI.h>
#include <Adb.h>

//The TMP36 output is connected on Arduino 0 analog port
int TMP36 = 0;

//temporary value for the analog reading
int reading;

// Adb connection.
Connection * connection;

// Elapsed time for ADC sampling
long lastTime;
```

```
// Event handler for the shell connection. In this case it
//will be empty since you do not accept any data from the
//phone
void adbEventHandler(Connection * connection, adb_eventType
event, uint16_t length, uint8_t * data)
{

}

void setup()
{

  // Note start time
  lastTime = millis();

  // Initialise the ADB subsystem.
  ADB::init();

  // Open an ADB stream to the phone's shell. Auto-reconnect
  connection = ADB::addConnection("tcp:4568", true,
  adbEventHandler);
}

void loop()
{

  //Check if 10 mins have elapsed since last transmission
  if ((millis() - lastTime) > 600000)
  {
    reading = analogRead(TMP36);

    // converting that reading to voltage, which is based
    //off the reference voltage
    float voltage = reading * 3.3;
    voltage /= 1024.0;

    // now convert to temperature in Celcius
    float temperatureC = (voltage - 0.5) * 100 ;
    //converting from 10 mv per degree wit 500 mV offset
    //to degrees ((volatge - 500mV) times 100)


    //For Fahrenheight conversion:
    //float temperatureF = (temperatureC * 9.0 / 5.0) +
    //32.0;

    //Send the temperature reading to the phone:
    uint16_t data = temperatureC;
    connection->write(2, (uint8_t*)&data);

    //Reset timer
    lastTime = millis();
```

```
    }

    // Poll the ADB subsystem.
    ADB::poll();
}
```

The Arduino Code – Code Review

You start the sketch code by including the essential libraries for the MicroBridge ADB:

```
#include <SPI.h>
#include <Adb.h>
```

You move on with defining the analog pin of the attached sensor (0) and the basic variable for sensor readings.

```
//The TMP36 output is connected on Arduino 0 analog port
int TMP36 = 0;

//temporary value for the analog reading
int reading;
```

The following line creates the ADB Connection object. It is actually a TCP Client object that will poll the server (running on the phone) for new events (data sent from the phone) and forward data to the phone as well.

```
// Adb connection.
Connection * connection;
```

You also use a long variable for checking time intervals between the ADB sampling (polling the server):

```
// Elapsed time for ADC sampling
long lastTime;
```

You then start with implementation the essential sketch methods. A very important function of the ADB library is the adbEventHandler. This function is invoked every data an event is received from the phone (i.e. some data is sent). For this case you do not need to implement something inside the function, since you do not expect any data from the phone. You still declare the function in the code though because is it needed for initializing the ADB connection in the setup() method.

```
// Event handler for the shell connection. In this case it
//will be empty since you do not accept any data from the
//phone
void adbEventHandler(Connection * connection, adb_eventType
event, uint16_t length, uint8_t * data) {}
```

Next, follows the default setup() method. You need to initialize the variables related to your sketch there. You have basically the time variable, which is set to the current time of Arduino system in milliseconds, and you also initialize the ADB, through the ADB::init() command.

```
void setup()
{

  // Note start time
  lastTime = millis();

  // Initialise the ADB subsystem.
  ADB::init();
```

In addition, you open an ADB (client) stream to the phone's shell. You use the same port you define the ADB server for listening on the Android code. Also, notice here that you need to refer the adbEventHandler method.

```
  connection = ADB::addConnection("tcp:4568", true,
adbEventHandler);
```

The loop() method will check if it is time to send data

```
void loop()
{

  //Check if 10 mins have elapsed since last transmission
  if ((millis() - lastTime) > 600000)
```

You read the temperature sensor and convert the input voltage to temperature according to the following formulas:

```
  reading = analogRead(TMP36);
```

You convert that reading to voltage, which is based off the reference voltage:

```
  float voltage = reading * 3.3;
  voltage /= 1024.0;
```

And then you convert to temperature in Celsius:

```
  //converting from 10 mv per degree wit 500 mV offset
  //to degrees ((volatge - 500mV) times 100)
  float temperatureC = (voltage - 0.5) * 100 ;
```

Finally, you send the temperature reading to the phone:

```
  uint16_t data = temperatureC;
  connection->write(2, (uint8_t*)&data);
```

and you reset the timer to the current time in milliseconds:

```
  lastTime = millis();
```

The Android Code

Due to size limitations, the code has not been included in the chapter. You can find the complete Android project in the book's code repository at **http://www.buildinginternetofthings.com.** The following section includes a full explanation of the code.

The Android Code – Code Review

For this project you need only to implement a single Activity class that will listen for incoming data from the Arduino and then will post a Twitter update with the temperature sensor reading.

You start by importing the appropriate Android libraries and the libraries for the ADB server implementation. In order to include the ADB Server library in your Android program, all you have to do is include the appropriate Server implementation source folder (folder named 'org') inside your project's source (src) folder. For a complete example you can check the source code of the ArduinoADB_Twitt project in the book's online code repository (**http://www.buildinginternetofthings.com**).

You also use an external library for Twitter communication, the twitter4j library that abstracts the authorization and post update complexity through simple method calls. You can download the latest version of the library from http://twitter4j.org/ and install the required library files (twitter4j-async-android-xxx.jar and twitter4j-core-android-xxx.jar) in your project's build path.

```
import java.io.IOException;
import org.microbridge.server.Server;
import org.microbridge.server.AbstractServerListener;
import twitter4j.Status;
import twitter4j.Twitter;
import twitter4j.TwitterException;
import twitter4j.TwitterFactory;
import twitter4j.conf.ConfigurationBuilder;
import android.app.Activity;
import android.os.Bundle;
import android.util.Log;
```

You then start writing the main activity class

```
public class ArduinoTwitt extends Activity {
```

You define a variable for receiving the sensor values:

```
private String SensorValue="0";
```

Then you create an ADB TCP server (based on MicroBridge LightWeight Server). This Server will run in a separate thread.

```
Server server = null;
```

You then implement the essential for the Activity onCreate() method. The method will instantiate the ADB Server and also add a listener to it for handling the incoming data:

```
/** Called when the activity is first created. */
@Override
public void onCreate(Bundle savedInstanceState) {
    super.onCreate(savedInstanceState);
    setContentView(R.layout.main);
```

Remember that you have to use the same port number used in ADK Main Board.

```
server = new Server(4568); //
server.start();
```

Once done with server instantiation, you add an event listener to the server. This listener (of AbstractServerListener type) includes the implementation of the onReceive() method (invoked every time you get an event – data from Arduino) which allows us to parse the data and send them to Twitter.

```
server.addListener(new AbstractServerListener() {
    @Override
    public void onReceive(org.microbridge.server.Client
    client, byte[] data){

        SensorValue = new String(data);
        TwittArduino("Arduino says it is
        "+SensorValue+"C
        in here!");

    }
});
```

You have implemented a separate method for Twitter communication. According to the twitter4j documentation you need to instantiate a ConfigurationBuilder object and set the consumer key, the consumer secret, the access token and the token secret. The latter are the information you get when you set the appropriate Twitter application (see Figure 6-10).

```
public void TwittArduino(String message) {
    ConfigurationBuilder cb = new
    ConfigurationBuilder();
    cb.setDebugEnabled(true)
    cb.setOAuthConsumerKey("XXXXXXX")
    cb.setOAuthConsumerSecret("XXXXXXX")
    cb.setOAuthAccessToken("XXXXXXX")
```

```
cb.setOAuthAccessTokenSecret("XXXXXXX");
```

The following TwitterFactory object (tf) will allow us to create a Twitter object and get an instance of it. Once done, it means that you have successfully connected to the Twitter application and you can post/get status updates.

```
TwitterFactory tf = new TwitterFactory(cb.build());
Twitter twitter = tf.getInstance();
```

Finally, you create a Status object so that you can update your Twitter status based on the sensor reading:

```
Status status;
    try {
        status = twitter.updateStatus(message);
        System.out.println("Successfully updated the
        status to [" + status.getText() + "].");

    } catch (TwitterException e) {
        // TODO Auto-generated catch block
        e.printStackTrace();
    }
```

Make sure you include the Internet permission in your AndroidManifest file. Once completed the code, just install the application and run it on your phone, after connecting the ADK board first. Then just watch your Twitter get updates based on the temperature of your room!

Project 3 - Control a relay switch by texting your phone

In the previous project you have experimented with the communication from the Arduino board towards the Android phone. In this project you will examine the event handler of ADB and use it in order the Arduino to receive commands from your Android phone. Android enables developers to interpret and parse incoming SMS messages from their own applications. So this gives us a great opportunity to control the outputs of your Arduino board through your phone by just using simple text messages.

In order to make something more useful than just lighting a led up, you will use a relay switch controlled by the Arduino. As explained in Chapter 2, a relay is actually a mechanical switch that is operated using electric current. It is used to control other circuits (mostly because you cannot handle the other circuit's current/voltage). It mainly consists of an electric magnet and an armature. When the electric magnet is activated (by current flowing within the relay when activated) the armature is moved, making or breaking this way an electrical connection activating or deactivating the circuit to control. When current is removed from the magnet, usually a spike of voltage is generated (due to the collapse of the magnetic field) that can make damage to the rest of your circuit. Therefore you should always use a diode when using relays in your projects. You can check the circuit setup for more details. With the relay switch you can control various electrical circuits and devices, like lights, heating, and basically anything that can be turned on using a switch.

The project consists obviously from two coding parts. The Arduino sketch that will listen from data from the Android device and control the relay, and the Android code that listens for incoming text messages and sends control flags through the ADB server. Let's take a look first at the project circuit.

The Circuit
For this project you will need the following:
- A relay switch
- A 2N3904 NPN transistor
- A 1N4001 diode
- A 1K Ohm Resistor
- An Arduino ADB Board
- An Android phone (version 1.5 and later) with a valid SIM card and know phone number
- (Optionally) A Breadboard for connecting the components with the Arduino

You will also need another phone to text the messages, or alternatively you can use an online SMS service for that purpose.

Consider the schematic of Figure 6-11. The relay you are using in this example has a coil voltage of 5V (DC), meaning the relay switch will be activated when 5VDC is supplied across the relay coil. The output pin of an Arduino does supply 5V, but the current that the relay requires to activate its switch is greater than the Arduino can safely supply. In this case, a 2N3904 NPN transistor is used to supply a higher current 5V source to the relay coil. The digital port 7 of Arduino controls the NPN transistor. By 'high' state it will close the circuit with the Gnd and allow current to flow the 5V source towards the relay switch. The relay switch will then change state and close the upper circuit.

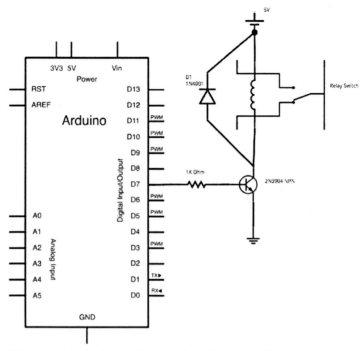

Figure 6-11. The circuit schematic for this project. The digital port 7 of Arduino controls the NPN transistor. By 'high' state it will close the circuit with the Gnd and allow current to flow the 5V source towards the relay switch. The relay switch will then change state and close the upper circuit.

An illustration of the circuit with a breadboard and an Arduino Uno is presented in the following figure. You need to replace the Uno with the Seeeduino ADK Main Board or use the USB Host Shield instead.

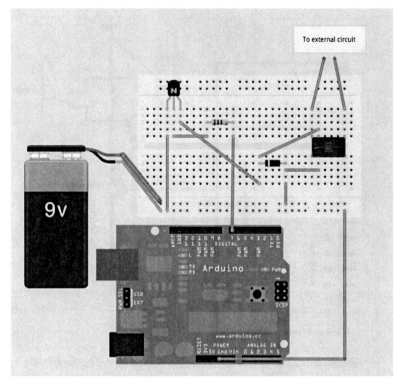

Figure 6-12. The circuit schematic for Project 2. The digital port 7 of Arduino controls the NPN transistor. By 'high' state it will close the circuit with the Gnd and allow current to flow the 9V source towards the relay switch. The relay switch will then change state and close the external circuit.

The Arduino Code

Listing 6-3. This Is the Arduino Code for receiving Events from the Android phone

```
#include <SPI.h>
#include <Adb.h>

//relay switch pin
#define RELAY_PIN 7

// Adb connection.
Connection * connection;

// Event handler for the shell connection.
void adbEventHandler(Connection * connection, adb_eventType
event, uint16_t length, uint8_t * data)
{
  int i;
```

```
  // Data packet contains one byte, the flag for turning on
  or off the relay
  if (event == ADB_CONNECTION_RECEIVE)
  {
    if(data[0]==1) {
      digitalWrite(RELAY_PIN, HIGH);
    }

    if(data[0]==0) {
      digitalWrite(RELAY_PIN, LOW);
    }

  }

}

void setup()
{

  // Initialise serial port
  Serial.begin(57600);

  //Set pin to output for controlling the relay
  pinMode(RELAY_PIN, OUTPUT);

  // Initialise the ADB subsystem.
  ADB::init();

  // Open an ADB stream to the phone's shell. Auto-reconnect
  connection = ADB::addConnection("tcp:4568", true,
  adbEventHandler);
}

void loop()
{

  // Poll the ADB subsystem.
  ADB::poll();

  delay(100);

}
```

The Arduino Code – Code Review

You start by including the appropriate libraries for the ADB communication.

```
#include <SPI.h>
#include <Adb.h>
```

You also define the Arduino digital pin where the relay switch is attached to:

```
#define RELAY_PIN 7
```

Then you create the ADB connection object:

```
Connection * connection;
// Event handler for the shell connection.
void adbEventHandler(Connection * connection, adb_eventType
event, uint16_t length, uint8_t * data)
{
  int i;

  // Data packet contains one byte, the flag for turning on
  or off the relay
  if (event == ADB_CONNECTION_RECEIVE)
  {
    if(data[0]==1) {
      digitalWrite(RELAY_PIN, HIGH);
    }

    if(data[0]==0) {
      digitalWrite(RELAY_PIN, LOW);
    }

  }

}

void setup()
{

  // Initialise serial port
  Serial.begin(57600);

  //Set pin to output for controlling the relay
  pinMode(RELAY_PIN, OUTPUT);

  // Initialise the ADB subsystem.
  ADB::init();

  // Open an ADB stream to the phone's shell. Auto-reconnect
  connection = ADB::addConnection("tcp:4568", true,
  adbEventHandler);
}

void loop()
{

  // Poll the ADB subsystem.
  ADB::poll();
```

```
delay(100);

}
```

The Android Code

Due to size limitations, the code has not been included in the chapter. You can find the complete Android project in the book's code repository at **http://www.buildinginternetofthings.com**. The following section includes a full explanation of the code.

The Android Code – Code Review

As usually you start with importing the appropriate libraries. First comes the MicroBridge library from which you use the Server object.

```
import org.microbridge.server.Server;
```

Then you need to import some Android-specific libraries that are essential for parsing the incoming SMS messages and libraries related to components you are using, like the Toast object for displaying short messages on the screen.

```
import android.content.BroadcastReceiver;
import android.content.Context;
import android.content.Intent;
import android.os.Bundle;
import android.telephony.SmsMessage;
import android.util.Log;
import android.widget.Toast;
```

For this Android application you need only one class that acts as a SMS listener and can have access to incoming events and content from the phone. This is done by extending the Android BroadcastReceiver class. Keep in mind that you do not create any interface for this app, meaning that when you install this app it automatically starts to listen for incoming SMS events.

```
public class SMSReceiver extends BroadcastReceiver
```

Inside the main class, you start by defining the global Server object.

```
Server server = null;
```

You continue by implementing the void onReceive() method. This is essential to be present in the code since the class extends the BroadcastReceiver class.

```
@Override
public void onReceive(Context context, Intent intent)
```

As the name suggests, this method is automatically invoked every time a SMS is received. So the following code retrieves the received SMS and parses the content. In addition to that it will check for a command (a 'true'/ 'false' flag) and send the respective flag to the Arduino through the ADB Server.

You start by getting a 'Bundle' object from the main activity (Android internal) that monitors for SMS events, which allows us to get the SMS message passed in.

```
Bundle bundle = intent.getExtras();
SmsMessage[] msgs = null;
String flag = "0";
```

In case there is some content, meaning that you do actually have a new SMS:

```
if (bundle != null)
```

You retrieve the SMS message received:

```
Object[] pdus = (Object[]) bundle.get("pdus");
```

In case more than one messages (in length) has been received you loop through all the messages and get the main message body and its content:

```
msgs = new SmsMessage[pdus.length];
//for each message get the main body content
for (int i=0; i<msgs.length; i++){
    msgs[i] =
    SmsMessage.createFromPdu((byte[])pdus[i]);
    flag = msgs[i].getMessageBody().toString();
```

The content is saved into the flag String variable. You check the flag for commands and communicate with the Arduino appropriately:

```
if(flag.equals("true") || flag.equals("false")) {
    try{
```

You initialize the ADB TCP server using the same port number used in ADK Main Board (4568):

```
server = new Server(4568);
server.start();
```

The Server starts at a different thread and you need to give it some time to startup before you send any data to the Arduino:

```
Thread.sleep(3000);
```

Then you send data in byte format. All you need to send is a '1' or '0' based on the incoming SMS flag. The Arduino will read and parse it according to the previous code listing.

```
byte data;

if(flag.equals("true")) data = 1;
else data = 0;
```

server.send(new byte[] {(byte) data});

Finally you stop the server. You make this so that in the next incoming SMS you can start the process again from the beginning of the onReceive method.

```
server.stop();
```

Summary

This chapter has presented to you all you need to connect your Arduino board with your Android phone and make them exchange data. We have discussed about hardware you need either when using Bluetooth or when connecting through USB. We have also covered the basic software libraries you will need for achieving the communication, like the Amarino toolkit or the MircroBridge ADB and Google ADK.

To demonstrate the functionality of connecting your Arduino with your phone, we have explored 3 projects: the first project sends air quality data to your phone using Bluetooth. The second project tweets sensor readings through your phone to your Twitter account and the third one demonstrates how to control Arduino from your phone using SMS messages.

Through the latter projects we have also examined briefly the communication of Arduino with the Internet using an Android phone. The next chapter presents more direct ways to communicate Arduino with the Internet and build IoT networks through wired and wireless interfaces

7 CONNECTING YOUR ARDUINO TO THE INTERNET

In order to communicate with the Cloud and build IoT projects you need access to the Internet. This Chapter will present you with several ways you can use in order to connect your Arduino directly or indirectly to the Internet. Directly refers to using an Internet-enabled communication module (such as a WiFi or an Ethernet Shield) whereas indirectly refers to using a gateway (such as your computer) for retrieving or forwarding information to the Internet.

You will examine both wired and wireless modules designed for the Arduino that enable Internet communication. The wired ones utilize Ethernet communication while for connecting wirelessly you can either use a WiFi (802.11x) module or a GSM mobile module. Examples with sketch codes and libraries are presented for each case. You will also learn how to create a Java application based on the Processing framework that communicates with the Arduino and provides Internet connectivity to your projects indirectly. Finally the Chapter discusses technologies and protocols specifically designed for machine2machine (M2M) communication and IoT networks, like the WebSockets and the MQTT protocol.

The Basics of the Internet

Let's start with overviewing the basic concepts behind the communication of a device (let it be your Arduino, PC, mobile phone, etc.) with the Internet. Although nowadays the connection to the Internet is pretty trivial for the user, in fact there are several (complex or less complex) mechanisms that take place when two devices are communicating with each other. The way these mechanisms handle the communication is described by protocols, like the TCP/IP that also defines how devices are identified using special network addresses (IP and MAC).

TCP/IP

One of the fundamental protocols that define communication between two Internet-enabled devices is the Transmission Control Protocol/Internet Protocol (TCP/IP). TCP/IP defines how the data should be coded, transmitted and received by the devices. It also describes how to devices initiate communication, when they shall exchange information and how the information is properly routed across various intermediate networks.

TCP/IP is often referred as a stack of layers because it actually consists of various layers each one implementing the connection, coding and transmission of the information mechanisms. The functionality TCP/IP stack is implemented within the operating system of the device. So when you are developing an application for a device you do not need to worry about such complex things. In the case of embedded devices like the Arduino (that they do not feature an operating system) the stack needs to be implemented by the hardware vendor that provides the appropriate network communication modules.

An alternate to TCP/IP is the UDP (User Datagram Protocol), which is less common in networking applications. The main difference between TCP/IP and UDP is that the latter uses a simple transmission mechanism without providing reliability, ordering, or data integrity. Thus, UDP provides an unreliable service and data packets may arrive out of order, appear duplicated, or even go missing. However UDP is quite faster protocol than TCP/IP (mainly because it avoid the data control mechanisms of TCP/IP) and is used in applications like voice streaming and online gaming.

IP/MAC Addresses

You probably have come across with Internet Protocol (IP) addresses when configuring your computer for joining a wired or wireless network. The IP addresses are 32-bit numbers, which translates into 4 groups of integers with range between 0-255 and are used to identify computers within a network (and the Internet of course). They are displayed in human-readable notations, such as 74.125.39.104 (one of Google search engine web servers).

Your PC, your mobile phone and yes your Arduino have one such address when they are connected on a wired and/or wireless network that gives them access to the Internet. Such addresses as the one mentioned previously belong to the IPv4 (Internet Protocol Version 4) and are still in use today. Due to the great demand for IP addresses (derived from the continuously increasing number of computing devices that access the Internet), a new version of the Internet Protocol is introduced (version 6 or IPv6) which features 128 bits for addressing. An IPv6 address looks like this: 2001:db8:0:1234:0:567:1:1. IP addresses are usually assigned by the network (by devices that control network access, like routers) to the connected devices. An IP address is unique within a network (no devices can have the same IP address) but a device can take different IP addresses each time it is connected to a network.

A Media Access Control (MAC) address is a unique identifier for the network modules that are pats of a network. They are used to identify devices and modules within the same network (i.e. computers that share the same Ethernet router or WiFi access point). Unlike IP addresses, a MAC address is usually assigned by the hardware vendor of the module and is always the same for each device no matter what the connected network is. When a device is trying to connect to a network it broadcasts its MAC address to other devices-members of the network so that the latter are aware of it. When the device is actually connected and an IP address is assigned to it by the router, the router makes an association between the assigned IP and the MAC address.

In Arduino, several networking module vendors do not provide the MAC address of their modules and there is also no direct way the Arduino itself to find out about the latter. Therefore it is essential to define a MAC address in your sketch code before connecting the Arduino to any wireless or wired network.

DNS and DHCP

The Domain Name Search (DNS) is a service that translates computer addresses like www.google.com (which are referred as URLs) to the respective IP address so that a device can connect to it through the Internet. The association of an IP address with a URL address like *www.buildinginternetofthings.com* is maintained in special Internet computers, called DNS servers, that act as search directories. In order your Internet device to be able to query and retrieve the IP address of a specific URL it needs to know the address of a DNS server and also be able to communicate with it using a specific protocol (usually provided by the operating system).

DHCP is a service that assigns IP Addresses to the devices members of a network. It is provided by the main network element (the router) as a service and also requires from the devices to be DHCP-capable.

In earlier versions of the Arduino, users had to manually set both MAC and IP address so that it could connect to a network. A code example of setting MAC and IP address using the Ethernet Library follows:

```
byte mac[] = {  0xDE, 0xAD, 0xBE, 0xEF, 0xFE, 0xED };
byte ip[] = { 192,168,1,100 };
Ethernet.begin(mac, ip);
```

The MAC address assigned programmatically (in the sketch) to the Ethernet module is DE:AD:BE:EF:FE:ED. You have to make sure that no other device in your network has such a MAC address. Then the begin() function of the Ethernet library is used to bind the MAC Address with the given IP.

In order to connect to a remote server, e.g., one of Google's Web servers, you had to use its IP address instead of its URL name:

```
byte server[] = { 173,194,33,104 };
Client client(server, 80);
```

Since Arduino v1.0, both DNS resolving and DHCP have been implemented within the Ethernet Library, so that the Arduino can both ask to be assigned of a network IP address and also lookup the IP address of a remote device:

```
byte mac[] = {  0x00, 0xAA, 0xBB, 0xCC, 0xDE, 0x02 };
Ethernet.begin(mac);
```

You can see that the begin() function needs only the MAC address. An IP will be required and automatically assigned by the network (in case there is no DHCP service available you can always set an IP manually).

In order to connect to a remote server you can do it by simply defining its URL name:

```
char serverName[] = "www.google.com";
EthernetClient client;
client.connect(serverName, 80)
```

Network Sockets

A Network Socket describes how two devices communicate using the TCP/IP protocol. More specifically it defines the ports that the communication will take place and the IP addresses (or URLs that are resolved into IP addresses) to be used. For example, if your computer wishes to contact the Google web page for a search, it will communicate with one of the Google's web servers' IP address on port 80 (which is the default port for web page communication). The Web server that listens on that IP address on port 80, has created a Server socket, whereas your browser creates a Client socket with the aforementioned IP, Port combination.

In programming working with network sockets is quite trivial when your application requires Internet connectivity or data exchange with remote computers. Even in cases you use a URL object, you might not define the port, since the default port 80 is used for HTTP connections. In all other cases you either create server or client sockets by defining both IP addresses and ports.

For instance, in Java a Client socket that connects to a Google Web server is constructed the following way:

```
import java.net.Socket;
...
Socket ClientSocket;
try {
ClientSocket = new Socket("74.125.39.105", 80);
}
catch (IOException e) {
System.out.println(e);
}
```

Usually the network library used (in the above case java.net.Socket) implements with the help of the operating system all the essential connections and handshaking between the two computers that the TCP/IP protocol defines. Fortunately, the majority of the IP-based communication modules that are available for the Arduino (either Ethernet or Wireless) implement such functionality. With the help of the appropriate libraries they make the creation of Network Sockets simple and straightforward:

```
byte server[] = { 173,194,33,104 }; // Google IP Address
Client client(server, 80);
client.connect();
...
```

Or in Arduino v1.0 and beyond:

```
char serverName[] = "www.google.com";
EthernetClient client;
client.connect(serverName, 80)
```

The code above is used to create client network sockets using the Arduino Ethernet Library. In a similar way, sockets are created when using an Arduino Wireless Library (more details on the following Sections).

HTTP

Connecting to a remote computer (e.g., a Web Server that can also be part of a Cloud system) using a Network Socket is one thing. Communicating properly with that computer and retrieving/sending information from and to it is a different thing. Like the TCP/IP requires a specific protocol to be followed by the devices for establishing a communication, the complete information exchange also requires that the devices follow a set of rules that describe how the connection is made, how the data is formatted and when information is transmitted. Such protocols are the FTP, SMTP, Telnet, HTTP, etc.

The HTTP is the Hyper Text Transfer Protocol that describes the communication for retrieving web content (e.g., web pages, web services, etc.) from a web server. It is considered a request-response protocol in the sense that for instance, a web browser acts as a client, while a web server (which is actually an application running on a computer hosting a web site) functions as a server. The client submits an HTTP request message to the server. The server, which stores content like HTML files, or performs other functions on behalf of the client, returns a response message to the client. Such a response contains completion status information about the request and in its message body any additional content requested by the client. The HTTP-based communication between the client and the server is made over the TCP/IP Protocol.

When the client requests communication and content retrieval from the server (i.e. when you ask web page from a web server) it initiates an HTTP session. An HTTP session is a sequence of network request-response transactions. An HTTP client initiates a request by establishing a connection to (usually) port 80 on the server. The server listens on that port and waits for a client's request message. Upon receiving the request, the server sends back a status line, such as "HTTP/1.1 200 OK", and a message of its own, the body of which is perhaps the requested resource, an error message, or some other information.

The request by the client is made using specially formatted messages. A request message contains the following:

- A Request line, such as GET /index.html HTTP/1.1, which requests a resource called index.html from a server
- Headers, such as Accept-Language: en or Host:www.google.com
- An empty line.
- An optional message body.

The latter data must be properly formatted when sent to the browser. For example, the request line and headers must all end with <CR><LF> (that is, a carriage return character like '\r' followed by a line feed character like '\n'). The empty line must consist of only <CR><LF> and no other whitespace.

When developing web-based applications you also come across with the 'request' methods of the HTTP. These are methods that indicate the desired action to be performed on a specific resource. The most common of them are the GET (requests a representation of the content of the specific resource, e.g., the content of a web page), the HEAD (similar to the GET but without the response body) and the POST request (which submits data to be processed, e.g., an HTML form). You will meet again the GET and POST requests in the various projects of this book that use them in order to send data from the Arduino (e.g., sensor readings) to a web-based application.

Let's examine next the various modules (wired and wireless) that can provide Internet connectivity to your Arduino!

Connect your Arduino using the Ethernet

The simplest way to provide Internet connectivity to your Arduino is by using the Ethernet protocol. It is considered simple because the respective Library is supported natively in the Arduino IDE, the additional hardware you need is the most affordable (compared to wireless solutions), it consumes less power (again compared to wireless connectivity) and it is the most common networking protocol you find in homes or small offices (you can simply connect it directly to your router using an Ethernet cable).

The Ethernet Protocol itself was on the first computer networking technologies introduced in 1980 for commercial usage. It has been used for creating Local Area Networks (LANs) and has become the dominant standard in home and office wired connectivity to the Internet. It is based on the idea of 'Shared Media', which initially was used to describe when computers were communicating over a shared cable acting as a broadcast transmission medium.

To connect computers with each other creating a LAN network, you need the appropriate interfaces on the computers (found in every commercial PC or laptop, referred as LAN or Ethernet ports), Ethernet cables and devices (called Hubs). When you need to interconnect two different networks (e.g., a LAN network connecting your PC and laptop and your Internet connection from your cable/DSL provider) you use a Router device. Routers also have additional features like assigning IP addressed to the devices automatically, etc.

What you will need

To make your Arduino part of a LAN network you need the appropriate Ethernet module that will provide you with a connection port and will implement whatever the Ethernet protocol defines for communicating with your Router and other networking devices.

The Arduino team has created an Arduino board that integrates an Ethernet module, called the Arduino Ethernet Board (http://arduino.cc/en/Main/ArduinoBoardEthernet). Alternatively you can use an Arduino Shield on your existing board like in Figure 7-1.

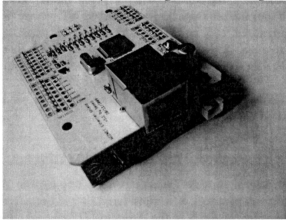

Figure 7-1. An Ethernet Shield on an Arduino clone (both produced by Seeedstudio)

When looking for an Ethernet Shield you need to pay attention to the Ethernet module used. Currently there are two modules used for Arduino Shields: The Wiznet W5100 Ethernet module and the Microchip's ENC28J60 Ethernet module. Only the first one is compatible with the official Arduino Ethernet Library (presented in the following Section). For the second module there is also a library available but it is quite different in the philosophy and usage within sketch code.

The Arduino Ethernet Library

The Arduino Ethernet Library is developed by the Arduino team and enables Arduino boards equipped with the appropriate controller to communicate with the Internet. You can use the library and create sketches that work either as a server accepting incoming connections (like a web server) or a client making connections to a server. The library supports up to four concurrent connections (i.e. incoming or outgoing or a combination of the latter).

The library comes preinstalled in your Arduino IDE. You can use it in your sketch through: Sketch->Import-> Ethernet. You will notice that the following headers will be added in the beginning of your sketch code:

```
#include <Dhcp.h>
#include <Dns.h>
#include <Ethernet.h>
#include <EthernetClient.h>
#include <EthernetServer.h>
#include <EthernetUdp.h>
#include <util.h>
```

The Dhcp.h header includes the essential functionality for requesting an IP address from the network router automatically. The Dns.h provides URL to IP address resolving and the Ethernet.h includes the basic Ethernet protocol functionality. Depending on your sketch application you can either use the EthernetClient.h or the EthernetServer.h for creating either a client that connects to a remote server or vice versa. The EthernetUdp.h provides implementation of the UDP protocol in case you need to create network connections that are less reliable but quite faster than TCP/IP (or because the server – application you need to connect to supports UDP only). The util.h contains some important variable definitions for the Ethernet Library.

When writing your own sketch for your Arduino Ethernet you do not need to include all these files. In most cases the Ethernet.h includes all the functionality you will need. However it is essential to include the SPI.h header file. The SPI (Serial Peripheral Interface) is used as a synchronous serial data protocol for communicating fast with the Ethernet module.

Let's see some examples that use the Ethernet Library and make your Arduino communicate with the Internet!

A Simple Ethernet Client Example

You will start by creating a client program that connects to a remote web server, which replies back with the IP address of your network router. The sketch reads the address and prints it over the Serial Monitor.

Upload the following code to your Ethernet-enabled Arduino:

Listing 7-1. The Simple Ethernet Client Sketch

```
#include <SPI.h>
#include <Ethernet.h>

// Enter a MAC address for your controller below.
byte mac[] = {  0xDE, 0xAD, 0xBE, 0xEF, 0xFE, 0xED };
//An IP Address and Router settings just in case there is no
DHCP available
IPAddress ip(192,168,1, 200);
IPAddress gateway(192,168,1, 1);
IPAddress subnet(255, 255, 255, 0);

//The Web Server you are connecting to
char serverName[] = "whatismyip.org";

// Initialize the Ethernet client library
EthernetClient client;

void setup() {
  // start the serial library:
  Serial.begin(9600);
  // start the Ethernet connection:
  if (Ethernet.begin(mac) == 0) {
    Serial.println("Failed to configure Ethernet using DHCP,
    trying to set IP mannualy");
    Ethernet.begin(mac, ip, gateway, subnet);
  }
  // give the Ethernet shield a second to initialize:
  delay(1000);
  Serial.println("connecting...");

  // if you get a connection, report back via serial:
  if (client.connect(serverName, 80)) {
    Serial.println("connected");
    // Make a HTTP request on the default page:
    client.println("GET / HTTP/1.0");
    client.println();
  }
  else {
    // If you didn't get a connection to the server:
    Serial.println("connection failed");
  }
}
```

```
void loop()
{
  // if there are incoming bytes available
  // from the server, read them and print them:
  while (client.available()) {
    char c = client.read();
    Serial.print(c);
  }

  // if the server's disconnected, stop the client:
  if (!client.connected()) {
    Serial.println();
    Serial.println("disconnecting.");
    client.stop();

    // do nothing forevermore:
    for(;;)
      ;
  }
}
```

Then connect an Ethernet cable (make sure it is connected to your Router and can provide Internet access) to the Ethernet port of your shield or Arduino Ethernet board and open the Serial Monitor on your Arduino IDE. You will see something like the following output:

```
connecting...
connected
HTTP/1.0 200 OK
Content-Type: text/plain
79.130.123.43
disconnecting.
```

A Simple Ethernet Client Example – Code Review
You begin the sketch by adding the essential Ethernet headers, i.e. the Ethernet.h and SPI.h

```
#include <SPI.h>
#include <Ethernet.h>
```

You enter manually a MAC Address for the Ethernet module (as mentioned before your Arduino is not capable of determining it on its own). Furthermore, in case your network router does not support automatic IP assigning, w also set valid IP addresses, the router's (gateway) IP address and a subnet mask using the appropriate variables.

```
byte mac[] = {  0xDE, 0xAD, 0xBE, 0xEF, 0xFE, 0xED };
```

```
IPAddress ip(192,168,1, 200);
IPAddress gateway(192,168,1, 1);
IPAddress subnet(255, 255, 255, 0);
```

Then you define the Web Server URL you are connecting to. In this case it is whatismyip.org that provides you with the IP address of your network router.

```
char serverName[] = "whatismyip.org";
```

Then you define the main Ethernet variable that will be used for making the appropriate network functions (connect to the network, connect to the server, etc.)

```
EthernetClient client;
```

The setup() function as usually performs all the appropriate initialization. You initialize the serial port and start an Ethernet connection to your network with the defined MAC address.

```
void setup() {
  // start the serial library:
  Serial.begin(9600);
  // start the Ethernet connection:
  if (Ethernet.begin(mac) == 0) {
    Serial.println("Failed to configure Ethernet using DHCP,
    trying to set IP mannualy");
    Ethernet.begin(mac, ip, gateway, subnet);
  }
```

In case the DHCP service is not available at the local router, the Ethernet will be initialized by the defined manual IP address. Like when configuring your computer, you also set the router's IP and subnet mask.

Next, you add a 1 second delay to give the Ethernet module some time to initialize properly. Then you proceed with the connection to the server URL address on port 80.

```
  delay(1000);

  if (client.connect(serverName, 80)) {
    client.println("GET / HTTP/1.0");
    client.println();
  }
```

You check for a successful connection (client.connect() will return true or false) and then send the request to the Web server. According to the HTTP protocol, you need to issue a GET request followed by the path of the resource (the default directory page indicated with '/' in this case) followed by the protocol version (HTTP/1.0). You also need to send a black line to tell the browser you have terminated the request and you are ready for feedback.

The server's response will be handled in the loop() function. The sketch checks for data from the server with the client.available() method and then reads each character arriving from the server using the client.read().

```
void loop() {
  while (client.available()) {
    char c = client.read();
    Serial.print(c);
  }
```

The following code lines just check if the server has disconnected, meaning that there is no other data sent and any transaction left and enters the sketch into an infinite loop so that there is no other interaction with the remote server.

```
  if (!client.connected()) {
    Serial.println();
    Serial.println("disconnecting.");
    client.stop();

    // do nothing forevermore:
    for(;;)
      ;
  }
}
```

A Simple Ethernet Server Example

Why would you implement a server on your Arduino instead of sending your data to the Internet? One good answer is that you might need an easy way to read sensor readings from your computing device (PC, laptop, smartphone, you name it), when you are at home without the need for an external Internet connection.

The best way to demonstrate this feature of the Arduino Ethernet Library is to use the default WebServer example provided by the Library. Open the sketch by selecting File->Examples->Ethernet->WebServer. This sketch implements a Web Server on your Arduino that prints the readings of the analog ports of your board into html format so that your browser can display them.

Make sure the default server address (192.168.1.177) is available for your local network or change it to something valid and upload the sketch to your Ethernet-enabled board. Once done, open your browser and enter the IP address defined in the sketch. You will see something like the following output:

analog input 0 is 320
analog input 1 is 294
analog input 2 is 279
analog input 3 is 277
analog input 4 is 276
analog input 5 is 280

Figure 7-2. The browser connected to the Arduino Web Server displaying the analog port values

Let's see the serve's code step by step. In the beginning, the essential Ethernet and SPI header files are includes, same as with the client example.

```
#include <SPI.h>
#include <Ethernet.h>
```

Similarly you define a MAC address for the Ethernet module.

```
byte mac[] = { 0xDE, 0xAD, 0xBE, 0xEF, 0xFE, 0xED };
```

The Web Server needs to have an IP address that you are aware of it in order to be able to access it through your browser. Therefore you do not request an IP address from the network using the DHCP service but you assign an IP address manually instead.

```
IPAddress ip(192,168,1, 177);
```

You then initialize the Ethernet server library defining the port you want to use (port 80 as default for the HTTP protocol):

```
EthernetServer server(80);
```

In the setup() function you bind the defined MAC address with the IP address. You also start the server for listening for connections using the internal Ethernet begin() function:

```
void setup()
{
  // start the Ethernet connection and the server:
  Ethernet.begin(mac, ip);
  server.begin();
}
```

The acceptance of client connections (i.e. your browser) is implemented in the loop() function. To handle connections and requests, an EthernetClient object is created. The client object is associated to the server instance (if successfully created in previous steps).

```
void loop()
```

```
{
   EthernetClient client = server.available();
   if (client) {
     boolean currentLineIsBlank = true;
```

The following loop is activated when a client is connected. While there is data received from the client it reads it and checks for a blank line (which according to the HTTP protocol means that the browser has completed its request). In such case, the server will reply with the default HTTP response header ("HTTP/1.1 200 OK") indicating that the request is properly received. It will also print a line declaring the content type that will be delivered to the browser ("text/html").

```
while (client.connected()) {
    if (client.available()) {
      char c = client.read();
       if (c == '\n' && currentLineIsBlank) {
        // send a standard http response header
        client.println("HTTP/1.1 200 OK");
        client.println("Content-Type: text/html");
        client.println();
```

Then the server prints to the browser the value of each analog input pin in a new line. To do so it prints on the end of each sensor reading the html new line character ("<br/>")."

```
   for (int analogChannel = 0; analogChannel < 6;
   analogChannel++) {
       client.print("analog input ");
       client.print(analogChannel);
       client.print(" is ");
       client.print(analogRead(analogChannel));
       client.println("<br />");
   }
```

Finally it gives the web browser time to receive the data and closes the connection.

```
   delay(1);
   client.stop();
```

Connect your Arduino using the WiFi

A more convenient method for connecting to the Internet is by using a WiFi network. It is considered more convenient since it requires no wiring and you can move around your Arduino, putting on places where no Ethernet cables are available (e.g., on a robot or carry it with you). However it is more energy-intensive so you need to be more careful with how often you send/receive data.

The WiFi (which comes from the term 'Wireless Fidelity') refers to the IEEE 802.11 family of standards. The latter is a set of standards for implementing wireless local area network (WLAN) communication between devices like PCs, laptops, mobile phones and any other device having an appropriate WiFi communication module. The most popular standards used are those defined by the 802.11b and 802.11g protocols. Due to the easy installation (no wiring and special infrastructure required other than a WiFi Access Point) and the relative low cost of the hardware needed, WiFi networks are the most popular for providing Internet connectivity in homes and small offices.

What you will need

In order to add WiFi connection abilities to your Arduino you need an appropriate WiFi module. The most common WiFi modules for the Arduino communicate with it through the Serial or SPI protocol and come us external shield modules, like the WiFi Bee from Seeedstudio or the WiFly Shield by Sparkfun (see Figure 7-3).

Figure 7-3. On the left: The WiFi Bee module compatible with XBee sockets (image courtesy of Seeedstudio). On the right: The WiFly Shield (image courtesy of Sparkfun).

Apart from the hardware you need libraries that take care programmatically of issues like communicating with the networking module over the serial interface or requesting an IP from the network router, implementing HTTP protocol requests, etc. Depending on the WiFi module you will use, the most common WiFi libraries utilized by all vendors are two: The WiShield Library by AsyncLabs and the WiFly Shield Library by Sparkfun.

The WiShield Library

The WiShield Library has been one of the initial Libraries developed for the Arduino platform for enabling Wireless communication that implements functionality like connecting to remote Web servers, hosting a Web Server on the Arduino and socket communication through a compatible Wireless

Shield – Arduino module. It can be obtained from https://github.com/asynclabs/WiShield and is mainly used by the WiFi Bee module from Seeedstudio.

Let's see an Internet connection example with the WiShield Library. You will use the same concept with the example presented for the Ethernet shield. You will request from www.whatismyip.org to send us back the IP address of our router.

Before plugging your WiFi Bee on your Arduino shield upload first the following sketch. Make sure first you have downloaded and installed the WiShield library in your Arduino IDE. In addition, you need to edit the sketch and set the proper values for the SSID variable with your Access Point's name, the security_type with the security type of your wireless network and the security_passphrase with the password for your wireless network.

Listing 7-2. An Example With The WiShield Library

```
#include <WiServer.h>
#define WIRELESS_MODE_INFRA 1
#define WIRELESS_MODE_ADHOC 2

// Wireless configuration parameters -----------
unsigned char local_ip[] = {192,168,1,2};    // IP address of
WiShield
unsigned char gateway_ip[] = {192,168,1,1}; // router or
gateway IP address
unsigned char subnet_mask[] = {255,255,255,0};  // subnet
mask for the local network
const prog_char ssid[] PROGMEM = {"YourWiFiNetworkName"};
   // max 32 bytes

//Modify according to your AP Security type
// 0 - open; 1 - WEP; 2 - WPA; 3 - WPA2
unsigned char security_type = 3;

// WPA/WPA2 passphrase
const prog_char security_passphrase[] PROGMEM =
{"YOURAPPASSWORD"}; // max 64 characters

// WEP 128-bit keys
// sample HEX keys
prog_uchar wep_keys[] PROGMEM = { 0x01, 0x02, 0x03, 0x04,
0x05, 0x06, 0x07, 0x08, 0x09, 0x0a, 0x0b, 0x0c, 0x0d,    //
Key 0
0x00, 0x00, 0x00, 0x00, 0x00, 0x00, 0x00, 0x00, 0x00, 0x00,
0x00, 0x00, 0x00,    // Key 1
0x00, 0x00, 0x00, 0x00, 0x00, 0x00, 0x00, 0x00, 0x00, 0x00,
0x00, 0x00, 0x00,    // Key 2
0x00, 0x00, 0x00, 0x00, 0x00, 0x00, 0x00, 0x00, 0x00, 0x00,
0x00, 0x00, 0x00    // Key 3
};
```

```
// setup the wireless mode
// infrastructure - connect to AP
// adhoc - connect to another WiFi device
unsigned char wireless_mode = WIRELESS_MODE_INFRA;

unsigned char ssid_len;
unsigned char security_passphrase_len;
// End of wireless configuration parameters ---------

// Function that prints data from the server
void printData(char* data, int len) {

  // Print the data returned by the server
  while (len-- > 0) {
    Serial.print(*(data++));
  }
}

// IP Address for www.whatismyip.org
uint8 ip[] = {98,207,221,49};

// A request that gets the IP of your Router
GETrequest getIP(ip, 80, "www.whatismyip.org", "/");

void setup() {
    // Initialize WiServer (you'll pass NULL for the page
    //serving function since you don't need to serve web
    //pages)
    WiServer.init(NULL);

    // Enable Serial output
    Serial.begin(57600);

    // Have the processData function called when data is
    //returned by the server
    getIP.setReturnFunc(printData);
}

void loop(){

  getIP.submit();

  // Run WiServer
  WiServer.server_task();

  //do nothing from here:
  for(;;) {}
}
```

Then plug the WiFi Bee into your shield and make sure it is connected to your computer via the USB cable. Open the Serial Monitor on your Arduino IDE and wait until the WiFi module initializes. Make sure your Serial Monitor is set to 57600 baud rate. It usually takes 5-10 seconds before the module is active and gets connected to the defined Wireless network. When done you will notice a blue LED on the shield turn on. Then check your Serial Monitor. You will notice that the response from the Web server is printed, including the HTTP headers and the IP address of your router.

The WiShield Library – Code Review
The sketch code begins with including the appropriate header file for the WiShield Library.

```
#include <WiServer.h>
```

You then define in variables the wireless access mode that can be used in your sketch. In general, the WiShield supports two modes: Infrastructure mode and AdHoc mode. The first one is used in case you are connecting to an existing wireless network provided by an Access Point or a Wireless router and the latter refers to when you need to connect directly to another WiFi-enabled device, like you laptop or your mobile phone. The first case is used since the Wireless router (or the Access Point) can provide Internet connectivity to your WiFi module.

```
#define WIRELESS_MODE_INFRA 1
#define WIRELESS_MODE_ADHOC 2
```

Next in the code comes the configuration of the wireless parameters. Here you need to define manually the IP address of your WiFi module (WiShield does not support DHCP), the IP of the Access Point or the Wireless router and the subnet mask (usually 255.255.255.0). You also define the name of your Wireless network.

```
unsigned char local_ip[] = {192,168,1,2};
unsigned char gateway_ip[] = {192,168,1,1};
unsigned char subnet_mask[] = {255,255,255,0};
const prog_char ssid[] PROGMEM = {"YourWiFiNetworkName"};
```

You also need to define the security type of your network, open, WEP, WPA or WPA2 using the numeric variable defined in the beginning of the sketch (0, 1, 2 or 3 respectively), and the password for your network.

```
unsigned char security_type = 3;
const prog_char security_passphrase[] PROGMEM =
{"YOURAPPASSWORD"};
```

The following lines define keys that will be used in case you select the WEP security type. These are the keys used for encrypting the data. You do not need to make any changes there, but make sure they are included in your sketch no matter what type of security you configure for your network.

```
// WEP 128-bit keys
// sample HEX keys
prog_uchar wep_keys[] PROGMEM = { 0x01, 0x02, 0x03, 0x04,
0x05, 0x06, 0x07, 0x08, 0x09, 0x0a, 0x0b, 0x0c, 0x0d,   //
Key 0
0x00, 0x00, 0x00, 0x00, 0x00, 0x00, 0x00, 0x00, 0x00, 0x00,
0x00, 0x00, 0x00,   // Key 1
0x00, 0x00, 0x00, 0x00, 0x00, 0x00, 0x00, 0x00, 0x00, 0x00,
0x00, 0x00, 0x00,   // Key 2
0x00, 0x00, 0x00, 0x00, 0x00, 0x00, 0x00, 0x00, 0x00, 0x00,
0x00, 0x00, 0x00    // Key 3
};
```

You then setup the wireless mode as infrastructure, since you are connecting to an Access Point.

```
unsigned char wireless_mode = WIRELESS_MODE_INFRA;
```

The following code segment contains the implementation of a function that handles the data from server. In this case while there is data available, you just print them to the Serial Monitor.

```
void printData(char* data, int len) {
    while (len-- > 0) {
    Serial.print(*(data++));
  }
}
```

You also define the IP Address for the server you are connecting to (www.whatismyip.org), since the WiShield library does not support DNS.

```
uint8 ip[] = {98,207,221,49};
```

You then create a GET request to the aforementioned IP address, on port 80, requesting the default page ('/'). You may notice that the WiShield library provides an implementation for GET requests, while the Arduino Ethernet Library requires from you to send the GET request as part of your communication with the server implementing what the HTTP protocol defines.

```
GETrequest getIP(ip, 80, "www.whatismyip.org", "/");
```

You are done with the appropriate variable and functions definition so you move on to the setup() and loop() functions. Inside the setup() function you need to initialize the WiFi module using the WiServer.init() method. In this case you pass NULL for the page serving function since you don't need to serve web pages. You also initialize the Serial output at 57600 bps and we also set to the GET request object you have created printData() function as return method. The latter means that when the GET request receives data from the web server, the data is handled by the defined function.

```
void setup() {
  WiServer.init(NULL);
  Serial.begin(57600);
  getIP.setReturnFunc(printData);
}
```

Finally, the loop() function activates the GET request by using the submit() method and also activates the server to perform its pending tasks (i.e. the GET request). It then enters an endless loop since you are not requiring any further interaction with the web server.

```
void loop(){
    getIP.submit();
    WiServer.server_task();

    //do nothing from here:
    for(;;) {}
}
```

The WiFly Shield Library
The WiFly Shield library is used a bit differently in your sketch code. First you need to obtain the latest version of the WiFly Shield Library from here: https://github.com/sparkfun/WiFly-Shield and install it on your Arduino IDE.

For the following example, open the example sketch 'WiFly_WebClient' in File->Examples->WiFlyShield and modify it according to the following.

Listing 7-3. An Example With The WiFly Shield Library
```
#include "WiFly.h"
#include "Credentials.h"

Client client("whatismyip.org", 80);

void setup() {
  Serial.begin(9600);
  WiFly.begin();

  if (!WiFly.join(ssid, passphrase)) {
    Serial.println("Association failed.");
    while (1) {
      // Hang on failure.
    }
  }

  Serial.println("connecting...");
```

```
  if (client.connect()) {
    Serial.println("connected");
    client.println("GET / HTTP/1.0");
    client.println();
  } else {
    Serial.println("connection failed");
  }
}

void loop() {
  if (client.available()) {
    char c = client.read();
    Serial.print(c);
  }

  if (!client.connected()) {
    Serial.println();
    Serial.println("disconnecting.");
    client.stop();
    for(;;)
      ;
  }
}
```

The Credentials.h file contains the following:

```
#ifndef __CREDENTIALS_H__
#define __CREDENTIALS_H__

// Wifi parameters
char passphrase[] = "passphrase";
char ssid[] = "ssid";

#endif
```

You will notice that the sketch looks very similar to when using the Arduino Ethernet Library and quite different from the WiShield Library. One major difference is that this sketch uses an external header file for defining the name (SSID) of the Wireless network and the access password. It then follows the philosophy of the Ethernet Library. You need to create a client object defining the URL and port you need to connect to.

In the setup() function you initialize the WiFi module and request to join the defined wireless network. Then you make the connection to the Web server using the connect() method and send the GET request to the server according to the HTTP protocol.

The loop() function receives any data sent from the Web server and prints them to the Serial Monitor. It then disconnects the client.

Connect your Arduino using a GSM Network

Another way of direct communication with the Internet is using a mobile (GSM) network. To do so you also need an appropriate GSM interface attached to your Arduino, like the GSM Shield in Figure 7-4 from Sparkfun. The idea is that you plug in a valid SIM card from your mobile operator that includes a data plan (so that you can access the Internet). The module communicates with the Arduino with a set of commands over a serial interface (like the Serial port). Depending on the GSM module vendor, the commands enable you to use the module for sending and receiving calls and SMS messages apart from connecting to the Internet using the TCP/IP.

This type of communication is much more flexible than the Ethernet and WiFi-based ones, since the cellular network coverage is much wider than WiFi making it a suitable solution especially when you need to build mobile projects or deploy your project on remote areas where no WiFi or Ethernet access is available.

Figure 7-4. The Cellular Shield for the Arduino by Sparkfun (image courtesy of Sparkfun).

However, cellular communication on the Arduino has some disadvantages: Firstly, it costs much more than a Wireless or Wired network and requires a data plan from your network operator. Secondly, the GSM modules are quite complicated in their function and communicate through commands using the serial protocol. The latter means that you have to implement the communication with the Internet using the command set provided by the vendor, which can be quite complex. For example, a connection to a web page using the module in Figure 7-4 would require the following command set:

```
+SIND:11 (register to network)
AT+CGDCONT=1,"IP","operator" (initialize TCP connectivity)
```

```
AT+CGPCO=0,"user","password", 1 (get authorized to the
network APN)
AT+CGACT=1,1  (activate Packet network)
AT+SDATACONF=1,"TCP","www.buildinginternetofthings",80 (open
TCP connection to remote host)
AT+SDATASTATUS=1 (query the socket status)
+SOCKSTATUS: 104 (104 means socket connected)
AT+SDATATSEND=1,13 (send data, 13 bytes)
>GET / HTTP1.1 <Ctrl+Z>
AT+SDATATREAD=1 (read incoming data)
. . .
```

Fortunately there is a GSM library for the Arduino available that simplifies the whole process. You can retrieve the library from: http://code.google.com/p/gsm-playground/

A sketch sample that uses functions instead of Serial commands for the previous communication example can be:

```
#include "GSM.h"
. . .
GSM gsm;
gsm.TurnOn();
gsm.InitGPRS("internet", "user", "password");
gsm.EnableGPRS(CLOSE_AND_REOPEN);
gsm.SendData("GET / HTTP/1.1\r\nHost:
www.buildinginternetofthings.com\r\n\r\n");
gsm.RcvData(20000, 1000, &ptr_to_data);
. . .
```

Other Ways to 'Internetize' your Arduino

What can you do in case you have not available an Ethernet or WiFi shield? Is your Arduino in this case isolated from the Internet world? The answer is fortunately no, since there are other ways to Internetize your Arduino. These are the 'indirect' ways for connecting an Arduino to the Internet since an intermediate, already Internet-enable device (like your mobile phone or your PC) can be utilized. In previous Chapter, among other examples, you were also presented with projects that forward information that comes from your Arduino to the Internet using your phone as a gateway. Still these projects require specific equipment (like an ADB enabled board or a Bluetooth modem) but can be a useful solution when there is no direct Ethernet or WiFi connection available to your Arduino.

A more simple solution requires only your USB cable and to make your PC as an Internet gateway for your Arduino. The Arduino communicates with your PC via the Serial port (as with the Serial Monitor) and with your own software you can send and receive data from the Arduino.

Using your PC as an Internet Gateway

By default your Arduino communicates with your Computer through the Arduino IDE for receiving your sketches and also for communicating (receiving input data or presenting information) with the Serial Monitor. The communication is performed through a Serial port. While the Arduino IDE does not provide any functionality for forwarding data to and from the Internet, you can easily develop your own application that makes so.

Depending on the development environment you will use, you will only need to have access and use the appropriate libraries for serial communication. For example, in Java you can use the Processing (http://processing.org) Library that also includes an implementation for serial communication.

Let's consider a simple example that demonstrates Internet connection through the PC: You will use your Arduino to control an output (e.g., a LED) depending on the weather forecast of your area. The temperature information will be retrieved from an online service using a Java application and forwarded to your Arduino. You will use the Google Weather API, which provides current weather conditions and short forecast information for the give city/place. It can be invoked like this: "http://www.google.com/ig/api?weather=Athens". The output is an XML-formatted file that contains information about the current weather status in Athens and a short prognosis. The output will look like this:

```
▼<xml_api_reply version="1">
  ▼<weather module_id="0" tab_id="0" mobile_row="0" mobile_zipped="1" row="0" section="0">
    ▼<forecast_information>
      <city data="Athens, Attica"/>
      <postal_code data="Athens""/>
      <latitude_e6 data=""/>
      <longitude_e6 data=""/>
      <forecast_date data="2011-11-17"/>
      <current_date_time data="2011-11-17 10:41:42 +0000"/>
      <unit_system data="US"/>
    </forecast_information>
    ▼<current_conditions>
      <condition data="Clear"/>
      <temp_f data="39"/>
      <temp_c data="4"/>
      <humidity data="Humidity: 69%"/>
      <icon data="/ig/images/weather/sunny.gif"/>
```

Figure 7-5. XML formatted response that contains weather information about Athens

For simplicity you do not want to use the whole information provided and thus parse the whole document, but only use the current temperature in order to turn on the LED in case the temperature exceeds some limits (e.g. is over 30 or lower than 15 degrees). In order to extract the temperature information you will parse the page and look for the temp_c or temp_f attributes that contain the current temperature information. Once the line that contains the attribute string is found, you simply retrieve the substring that contains only the temperature value.

You will be presented with two different methods for implementing the latter example, one reading and writing directly to the Serial Port and one using the Arduino Library for Processing.

The Java - Processing Code

For the first implementation example you will need to have the Java Development Toolkit (JDK) installed and preferably a Java IDE (e.g., Eclipse) for development. You will also need to install the essential Processing Library files (core.jar, serial.jar and RXTXcomm.jar) in your system's and/or development environment path.

Processing has been initially created for developing graphical applications. In order to utilize Processing in your code you have to create a class that extends the PApplet Processing class. It also follows the same coding philosophy Arduino sketches do (one would expect so since Arduino IDE is based on Processing), meaning that there are two main functions that are invoked during the execution: one for initializing variables and objects and one that is continuously executed when the application is running.

After you have properly set up your development environment, create a Java class named 'Gateway' with the following code:

Listing 7-4. The Java Processing Code For Communicating With The Arduino Over The Serial Port

```java
import java.io.BufferedReader;
import java.io.InputStreamReader;
import java.net.URL;

import processing.core.PApplet;
import processing.serial.Serial;

public class Gateway extends PApplet {
    Serial myPort;
    String inString;
    String temperature;
    String[] tmp;

    public void setup () {

        // List all the available serial ports
        println(Serial.list());

        // Open whatever port is the one you're using.
        //Usually it is the
        //first port in the list
        myPort = new Serial(this, Serial.list()[0], 9600);
```

```
        // Do not generate a serialEvent() unless you get
        a newline character:
        myPort.bufferUntil('\n');
}

public void draw () {

    //Read strings from the Serial port until new line
    //character
    inString = myPort.readStringUntil('\n');
    if(inString != null) {

        if(inString.equals("weather")) {
        try {
            URL url = new
            URL("http://www.google.com/ig/api?weather=At
            hens");

            BufferedReader in = new BufferedReader(new
            InputStreamReader(url.openStream()));

            String inputLine;
            while ((inputLine = in.readLine()) !=
            null)
            {

                int p = inputLine.indexOf("temp_c");
                if(p>0) {
                    //Get a substring of the line
                    //containing the temperature
                    temperature =
                    inputLine.substring(p, p+16);
                    //Split the string into parts
                    //based on '"'
                    tmp = temperature.split("\"");
                    //The second part will contain the
                    //plain temperature value
                    temperature = tmp[1];

                    myPort.write(temperature);
                    break;
                }
            }

            in.close();
        } catch (Exception e) {
            e.printStackTrace();
        }
    }
  }
}
```

```
//Do not need to implement anything here
// The PApplet will execute any code inside the setup()
//and draw() functions
public static void main(String[] args) {

    }
}
```

The code above requires that you have connected your Arduino board and that you have properly set up your Java environment importing the Processing Library jar files. It also requires an appropriate sketch to be executed on your Arduino that will send the essential String flag over the Serial port and retrieve the temperature value. Before moving on with the sketch code let's take a closer look at the Java code.

The Java - Processing Code – Code Review
The code starts by importing the essential libraries for the communication with the Internet and the Arduino through the Serial Port:

```
import java.io.BufferedReader;
import java.io.InputStreamReader;
import java.net.URL;
import processing.core.PApplet;
import processing.serial.Serial;
```

Then you define the main program class (Gateway) that will be executed as a Processing Applet and therefore needs to extend the PApplet class:

```
public class Gateway extends PApplet {
```

Inside the class you define some global variables, the first one is the serial port for the communication with the Arduino and the second one is a String variable used to parse incoming data from the Arduino (i.e. the 'weather' message that will initiate acquiring the weather status).

```
Serial myPort;
String inString;
String temperature;
String[] tmp;
```

As discussed, a Processing Applet is structured the same way an Arduino sketch does. It consists of two main functions, the setup() function that initializes variables and the draw() function that is continuously invoked (like the Arduino loop() function).

In the setup() function you print all the available serial ports (mostly used for debugging in case your Arduino is not connected in the first one available) and then initialize the Serial port object (myPort) by assigning the first serial port to it and the proper serial baud rate (same that will be defined in the Arduino sketch). The setup() function also instructs the Serial port object to buffer the incoming data until a new line character ('\n') is received (i.e. a full message has been received).

```
public void setup () {
        // List all the available serial ports
        println(Serial.list());

        // Open whatever port is the one you're using.
        //Usually it is the
        //first port in the list
        myPort = new Serial(this, Serial.list()[0], 9600);
        // Do not generate a serialEvent() unless you get
        //a newline character:
        myPort.bufferUntil('\n');
}
```

The draw() function will listen for incoming data from the Arduino through the Serial port. It will check whether the Arduino requests for a new weather update (i.e. has sent a 'weather' flag string in the Serial port) and will request the output of the Google Weather API.

```
public void draw () {

        inString = myPort.readStringUntil('\n');
        if(inString != null) {

        System.out.println(inString);
        if(inString.equals("weather")) {
        try {
            URL url = new
            URL("http://www.google.com/ig/api?we
            ather=Athens");
```

In order to read the output of the URL you need to create an InputStreamReader object connected to the URL stream. For better parsing of the InputStreamReader you utilize a BufferedReader object:

```
BufferedReader in = new BufferedReader(new
InputStreamReader(url.openStream()));
```

By the code line above, the URL connection and web page request has been made and the BufferedReader object is ready to receive data from the Web Server.

You use a String object to read the output of the BufferedReader object

```
String inputLine;
```

You read the web page data as served by the Webserver in lines and assign it to the inputLine String variable. While the arrived input data is not empty you look for the 'temp_c' string attribute so that you can retrieve the temperature value.

```
while ((inputLine = in.readLine()) != null) {
    int p = inputLine.indexOf("temp_c");
```

To search for "temp_c" in the String you use the indexOf method. If it returns a value greater than zero then the attribute has been found and you need to process the string further to retrieve the temperature value. To do so you just split the inputLine string from the point the attribute has been found (assigned into the p variable) to p+16 characters (p+16 is a safe length that contains the temperature value). The you split the resulting substring into parts based on the "" delimiter. The second part will contain the plain temperature value.

```
if(p>0) {
    temperature = inputLine.substring(p, p+16);
    tmp = temperature.split("\"");
    temperature = tmp[1];

    myPort.write(temperature);
    break;
}
```

In case you need to retrieve the value in Fahrenheit (instead of Celsius) you just need to replace the "temp_c" with "temp_f".

Using an XML parser for that is much more efficient but for simplicity you can always use the String parsing functions.

Finally, you close the BufferedReader stream:

```
in.close();
```

The main() function does not need to have anything implemented. The PApplet will execute any code inside the setup() and draw() functions.

The Arduino Code

The Arduino sketch needs to send a request (string 'weather') to the Serial Port, read the temperature and turn on/off the LED based on the temperature value.

Listing 7-5. The Arduino Sketch For Communicating with the Java Application Over the Serial Port

```
//Helpful variables for reading from Serial Port
char InString[8];
int sb;
int  counter  = 0;

void setup() {
```

```
// initialize the digital pin as an output.
// Pin 13 has an LED connected on most Arduino boards:
pinMode(13, OUTPUT);

// start serial port at 9600 bps:
Serial.begin(9600);
}

void loop() {
    //Ask for a weather update from the Java Application
    Serial.println("weather");
    //read incoming data:
    if(Serial.available()) {
      while (Serial.available()){
          sb = Serial.read();
          InString[counter] = sb;
          counter++;
      }
      int temp = atoi(InString);

      if(temp>30 || temp < 15) {
          digitalWrite(13, HIGH);
       }
      else {
       digitalWrite(13, LOW);
      }
   }

  //check every 10 seconds for new temperature forecast:
  delay(10000);
}
```

The Arduino Code – Code Review
The Sketch starts by defining helpful variables for reading from Serial Port:

```
char InString[8];
int sb;
int  counter  = 0;
```

The setup() function initializes the pin 13 as an output PIN so that the Arduino LED can be turned on/off. It also starts the Serial Port communication at 9600bps.

```
void setup() {
  // initialize the digital pin as an output.
  // Pin 13 has an LED connected on most Arduino boards:
  pinMode(13, OUTPUT);

  // start serial port at 9600 bps:
  Serial.begin(9600);
}
```

Inside the loop function you implement the main communication with the Java Application. Initially the sketch will send the 'weather' command over the Serial Port so that the Application can retrieve the temperature information. Then it will listen for incoming data through the Serial Port.

While the Serial data arrives, it reads each character and stores it into the global char array (InString) variable. The counter variable (as the name suggests) counts the received characters and increments the array position appropriately.

```
void loop() {

    //Ask for a weather update from the Java Application
    Serial.println("weather");
    //read incoming data:
    if(Serial.available()) {
        while (Serial.available()){
            sb = Serial.read();
            InString[counter] = sb;
            counter++;
        }

    }
```

Once the while loop has finished, meaning that all characters have been received from the Java Application, it converts the temperature value from char format into integer using the standard C atoi() function:

```
int temp = atoi(InString);
```

Finally, it checks whether the temperature is between a specific range and turns on/off the LED.

```
if(temp>30 || temp < 15) {
    digitalWrite(13, HIGH);
}

else {
    digitalWrite(13, LOW);
}
```

At the end of the loop function you add a 10 second delay before the sketch asks again for temperature data through sending the 'weather' command.

```
//check every 10 seconds for new temperature forecast:
delay(10000);
```

As you may have noticed, you need to develop both the Java Application code and the Sketch code appropriately so that they can both communicate, by defining how each program will interact with the other: you need to define both the data format (in this example, the value of the temperature) and when data communication shall be initiated (in this example, when the Arduino sketch sends the 'weather' command). More or less, it is like defining your own custom communication protocol!

An easier way is to use the Arduino Library for Processing as explained in the following section.

Using the Arduino Library for Processing

The Arduino team has created a Processing library that allows you to control your Arduino board directly from the Processing environment or from a Java Application that uses the Processing Library without writing a sketch for the Arduino. Instead, you need upload a standard sketch to the board and communicate with it using the library. The sketch is called Firmata, and is included in the Arduino IDE in the examples section.

The functionality described in the previous example can be implemented entirely on the Java Application, since the latter will both acquire the temperature information from the Internet and control the Arduino output.

In order to use the Arduino Library for Processing in your own applications, you need to install the respective jar file to your programming environment and/or your Java runtime environment. You can download the latest library version from here: http://www.arduino.cc/playground/Interfacing/Processing

The Java Code

You only need to create a Java Class like the following one:

Listing 7-6. The Java Code For Communicating with the Arduino using the Arduino Library for Processing

```
import java.io.BufferedReader;
import java.io.InputStreamReader;
import java.net.URL;

import processing.serial.*;
import cc.arduino.*;
import processing.core.PApplet;


public class Gateway_ArduinoLib extends PApplet{
    //The Arduino Object
    Arduino arduino;
    //The LED pin
    int ledPin = 13;
    String temp;

    public void setup(){
        //Initialize the Arduino object connected on the first
        //available Serial Port
        arduino = new Arduino(this, Arduino.list()[0], 57600);
```

```
    //Define the LED pin as OUTPUT
    arduino.pinMode(ledPin, Arduino.OUTPUT);
}

public void draw(){

    try {
        URL url = new
        URL("http://www.google.com/ig/api?weather=Athens
    ");
        BufferedReader in = new BufferedReader(new
        InputStreamReader(url.openStream()));

        String inputLine;
        while ((inputLine = in.readLine()) != null) {
            int p = inputLine.indexOf("temp_c");
            if(p>0) {

                System.out.println(inputLine.substri
                ng(p+13, p+14));
                temp = inputLine.substring(p+13,
                p+14);
              break;
            }
        }

        in.close();

    } catch (Exception e) {
        e.printStackTrace();
    }
    if(Integer.parseInt(temp)>30 ||
        Integer.parseInt(temp) < 15) {
        arduino.digitalWrite(ledPin, Arduino.HIGH);
    }
    else {
        arduino.digitalWrite(ledPin, Arduino.LOW);
    }

    //check every 10 minutes for new temperature
    //forecast:
    delay(600000);

}

//Do not need to implement anything here
// The PApplet will execute any code inside the setup()
//and draw() functions
public static void main(String[] args) {

}
}
```

Before running the Java code, open your Arduino IDE and upload the StandardFirmata Sketch (File -> Examples -> Firmata -> StandardFirmata). When done, run the code and watch the Arduino turn on/off the LED.

Let's see how the code works.

The Java Code – Code Review

As with the previous Java Application you start by importing the appropriate libraries for retrieving the web page.

```
import java.io.BufferedReader;
import java.io.InputStreamReader;
import java.net.URL;
```

You also import the Processing libraries that enable the Serial communication. Notice that in this case, you also import the Arduino Library for Processing (cc.arduino).

```
import processing.serial.*;
import cc.arduino.*;
import processing.core.PApplet;
```

You define the same class that also extends the PApplet class (since you are using Processing).

```
public class Gateway_ArduinoLib extends PApplet{
```

Inside the class you define some global variables, like the Arduino object the led pins (the same way you would in an Arduino sketch) and a String variable for handling the temperature value.

```
    Arduino arduino;
    int ledPin = 13;
    String temp;
```

Similarly, you use the setup() function for initializing the Arduino object by defining the Serial port that the board is connected to (usually the first one, so Arduino.list()[0] is used) and you also set the digital LED pin to output mode (exactly as you would do in an Arduino sketch).

```
    public void setup(){
      arduino = new Arduino(this, Arduino.list()[0], 57600);
      arduino.pinMode(ledPin, Arduino.OUTPUT);
    }
```

The draw() function as in previous example, will connect to the weather service, parse the textual output and detect the temperature value and will instruct the Arduino to turn on/off the LED. The main differences with the previous example are two: a) in this case the Java code does not receive a command from the Arduino and b) the control of the digital pin is made through the library and the Arduino object instead of sending the temperature value over the Serial Port. The internal communication between the Java application and the Arduino itself is implemented by the library and you do not need to worry about it.

```
try {
    URL url = new
    URL("http://www.google.com/ig/api?weather=Athens
");

    BufferedReader in = new BufferedReader(new
    InputStreamReader(url.openStream()));
        String inputLine;
        while ((inputLine = in.readLine()) != null) {

            int p = inputLine.indexOf("temp_c");
            if(p>0) {

                System.out.println(inputLine.substring(p
                +13, p+14));
                temp = inputLine.substring(p+13, p+14);
                break;
            }
        }

        in.close();

} catch (Exception e) {
    e.printStackTrace();
}
if(Integer.parseInt(temp)>30 ||
Integer.parseInt(temp) < 15) {
    arduino.digitalWrite(ledPin, Arduino.HIGH);
}
else {
    arduino.digitalWrite(ledPin, Arduino.LOW);
}

//check every 10 minutes for new temperature
//forecast:
delay(600000);

}
```

The delay is introduced at the end of the draw() function, the same way you did with the Arduino sketch in the previous example. The method for turning on/off the LED is arduino.digitalWrite(ledPin, Arduino.HIGH) and arduino.digitalWrite(ledPin, Arduino.LOW) respectively.

The Publish/Subscribe Notion

The basic functionality of devices like the Arduino that belong to an IoT network is to sense and send data to Internet-based services or to receive data and commands from the latter and perform various actions.

As you have seen in this Chapter (and you will also notice in the following Chapters of the Book) communication is performed over the HTTP protocol since most of the Cloud-based services for managing sensor data (Cosm, Nimbits, etc.) provide HTTP-based interfaces (APIs) for communicating with the Arduinos and other sensor devices. According to the HTTP specification, each time a request is made by the Arduino (client) to a web application (server) the client initiates an HTTP session, makes a GET or POST request with the appropriate header information, waits for response from the server and closes the connection. It is obvious that there is a lot of unnecessary data transmitted and too many connections opened and closed in case you wish to send data quite frequently (e.g., every second).

In addition, the client has to make a connection any time it expects some input from the server (e.g., a command for activating a digital output). This means that if you want to develop a project that controls the outputs of your Arduino based on commands/instructions from the Internet, you also need to implement a mechanism on the Arduino that will periodically check for commands on a remote server. Alternatively you could implement the server on the Arduino, but this introduces other issues like configuring port forwarding on your Internet router and it is also quite difficult to implement in case you need to control more than one Arduinos.

Since embedded systems are becoming enhanced with networking capabilities and the Internet of Things becomes a reality, new protocols and ways of communication are proposed that are addressing the aforementioned issues.

Web Sockets

The Web Sockets is actually what the name suggests: Network sockets that run over the HTTP protocol, used mostly within Web Browsers. They have been introduced by the new HTML5 standard as an application communication interface. Using Web Sockets you can have two Internet-enabled devices can communicate through a full-duplex communications link that operates over a single network socket. Initially they have been designed for usage within HTML5-compliant browsers using JavaScript.

Web Sockets are the typical example of the Publish/Subscribe notion. Their great advantage is (since they are sockets) that you can make a connection from a client (e.g., your Web browser) to a Web server and retain that connection open without having the server terminate it every time it delivers some information to you (as the original HTTP protocol defines). It is like subscribing to a communication channel. Whenever there is some information to receive (e.g., a sensor update) you can receive from the server without the need to poll it frequently. Similarly, if your client needs to send information, it just does so without the need to initiate a new connection to the Server.

Using the Web Socket interface is quite easy: To connect to a Server that offers such functionality, you just need create a new Web Socket instance, providing the new object with a URL that represents the end-point URL to which you wish to connect. In JavaScript it is coded as follows:

```
var myWebSocket = new WebSocket("ws://www.websocket.org");
```

The 'ws://' indicates that you are using a Web Socket connection. Such a connection is established by upgrading from the HTTP protocol to the Web Socket protocol during the initial interaction between the client and the server. Once established, Web Socket data frames can be sent back and forth between the client and the server in full-duplex mode. The connection itself is exposed via the onmessage and postMessage methods defined by the Web Socket interface.

Before connecting to a remote server and sending/receiving messages, you can associate a series of event listeners to handle each interaction:

```
myWebSocket.onopen = function(evt) {
    alert("Connection open ..."); };
myWebSocket.onmessage = function(evt) {
    alert( "Received Message:  "  +  evt.data); };
myWebSocket.onclose = function(evt) {
    alert("Connection closed."); };
```

To send a message to the server, simply call postMessage and provide the content you wish to deliver. After sending the message, use the disconnect() function to terminate the connection.

```
myWebSocket.postMessage("Hello Web Socket! I am an Arduino
fan!");
myWebSocket.disconnect();
```

To see an example of Web Socket communication on your own browser, you can visit http://websocket.org/echo.html, follow the instructions and also see the code needed for implementing your own example.

A very nice and open-source implementation of a Web Socket-based server is provided for the Arduino by Per Ejeklint. You can find the library here: https://github.com/ejeklint/Arduino-Websocket-Server. According to the developer, the implementation is still quite draft and limited, but it is enough to give you an idea about Web Sockets and how to use them on your Arduino.

Implemented for the Arduino Ethernet Library it creates a Web Socket server that listens for incoming connections. You can then handle the received data. For demonstration purposes there is also an HTML file available that implements the Web Socket client on your browser.

The MQTT protocol

MQTT is the Message Queue Telemetry Transport protocol introduced several years ago as an open message protocol that enables the transfer of data like sensor readings in the form of messages from embedded devices to remote servers.

The last version of the protocol (MQTT v3.1) provides a Publish/Subscribe messaging model that is implemented in a very lightweight way. It is thus considered very useful for connections with remote locations using embedded devices with limited resources.

The Facebook Messenger mechanism for the iPhone and Android Apps is based on the MQTT protocol for distributing chat messages.

MQTT, compared to Web Sockets, it is not based on the HTTP protocol so it cannot be executed within a browser and requires a separate implementation for both client and server applications. There are a few free/open-source MQTT server implementations available, like the RSMB (http://www.alphaworks.ibm.com/tech/rsmb) and the Mosquitto Server (http://mosquitto.org/).

More information on MQTT and available implementations can be found in http://mqtt.org.

There is also a MQTT Client Library implementation for the Arduino by Nick O'Leary. It can be found here: http://knolleary.net/arduino-client-for-mqtt/ and currently it supports both the Arduino Ethernet Shield and the WiFly Shield by Sparkfun. In order to use it you will need to set up your own MQTT server first and then utilize it for subscribing your Arduino board and sending/receiving messages.

Send Arduino Data to your own Cloud Application

It is pretty obvious that your Arduino can communicate by several means with the Internet and exchange information with remote computers. Some of the projects you might be interested in will have to do with retrieving information from the Internet (such as weather information) so that your Arduino can control for example some actuators or simply display the information on a LCD screen. But what if you want to use an Internet Service for sharing your sensor readings on line? One solution is to use existing Web applications (the following chapters will demonstrate how to use some of the most common Web-based services for managing sensor information, like the Cosm and the Nimbits) and the other one is to create your own web application!

Let's see how you can build your own Cloud-hosted application that will receive your sensor readings directly from your own Arduino!

Project Description

The idea of this project is to develop a web-based application that will be hosted on a Cloud infrastructure. The application will allow you to store sensor readings from your Arduino and also to visualize them so that you can view them anywhere, anytime using a Web browser and an Internet connection.

To develop and deploy such an application we need the following:

- A Cloud service that will host the application and will provide all the essential components for the proper functionality of the latter (like an appropriate web or application server, a database server, etc.).
- The Web application itself that will provide a lightweight interface for the Arduino to send data over HTTP (through a GET or POST request)
- An Internet-enabled Arduino, either using a WiFi module or an Ethernet Shield/Board
- The Sketch code that will read analog/digital ports and send readings to the Web application on the Cloud.

To meet the first requirement you will utilize the Google App Engine for hosting the Web application and an Ethernet-enabled Arduino for communicating with the latter. Based on the presented sketch you can easily create a new one that uses a WiFi module for connecting to the Internet.

Application Features

The application will read GET requests from the client (the Arduino). The requests will contain the sensors' name (so that it is feasible to use the same application for different sensors on the Arduino) and the current value of the sensor. The value will be stored into the datastore provided by the Cloud service so that a history of sensor readings can be later retrieved. You will use Java to develop a Servlet for the application.

The application should be also able to visualize (on a browser) the data history of a particular sensor. Therefore a second application will be developed and deployed that will receive the name of a particular sensor through a GET request (from your browser) and display a graph with the last 10 sensor readings. The Google Chart API will be also used for the visualization of the graph.

The Google App Engine

The Google App Engine is a Cloud computing platform for developing and hosting web applications in Google's infrastructure. It is considered a Cloud-based platform because of the following features:

- It offers automatic scaling for web applications as the number of web requests increases for an application,
- It automatically allocates more resources for the web application to handle the additional demand.
- It does not require from the developer to set up an application server and a database for the needs of the application.

One major benefit of using the Google App Engine is that it is free up to a certain level of utilized resources. Fees are charged for additional storage, bandwidth, or CPU usage required by the application. For developing and testing your own application on a Cloud environment it is a very low-cost and easy solution. It provides a local development environment that simulates Google App Engine on your computer and a very easy mechanism to deploy your application on the Cloud with the click of a button. It also provides an interface for authenticating users and sending emails using Google accounts.

Regarding data storage it provides a distributed storage service that features a query engine and transactions. The latter is not like a traditional relational database (e.g., like MySQL). Data instead of tables are stored into objects, or "entities", that have a kind and a set of properties. You can still make SQL-like queries that can retrieve entities filtered and sorted by the values of given properties.

Currently, the supported programming languages are Python, Java, and Google's scripting language Go. More information on Google App Engine can be found on the official web site at: http://code.google.com/appengine.

Let's move on with the development of you own Application!

Building a Java Web App for Google App Engine

You will start with the Web-based application that will be hosted on the Google App Engine. Based on the availability of the supported programming languages, the application will be developed using Java. For creating a web-based application on Java, you need to build a Servlet. A Servlet is a specific Java class that extends the capabilities of the hosting server (called application server) by accepting and responding to requests from clients (browsers or other applications). The application hosting server of a Servlet is actually a Web server that has also a special component, called Web container that is capable of managing and interacting with the Servlet.

Servlets typically embed HTML inside Java code, which translates to that when the Servlet code will be executed on the Web container, it will generate the appropriate HTML code for the client.

For this particular example you will also develop and use a JSP page (referred previously as the visualization web application). JavaServer Pages (JSP) is another Java technology that provides Web server the ability to serve dynamically generated web pages based on HTML (and other formats). In contrast to Servlets, JSPs embed Java code in HTML.

More information on how Servlets and JSPs work and development instructions can be found on any of the book series about Java Web development (like 'Beginning Java EE 6 with GlassFish 3' and 'Beginning JSP 2').

The process of development and deployment of the Web application requires three basic steps: a) to create a new application on the Google App Engine environment, b) to set up properly the Eclipse development Environment and c) to develop and deploy the Java code on the Engine.

Create an Application on the Google App Engine

The first step is to create a new application on the Google App Engine. To do so you need to create (or already have) a Google account and then visit: https://appengine.google.com

You may be asked to enter your account credentials again and/or provide some complimentary information like your phone number. After successfully logged in your will be welcomed and prompted to create a new Application as illustrated in Figure 7-6.

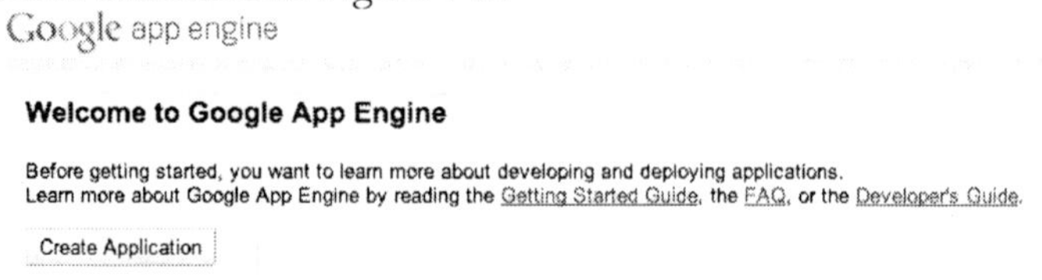

Google app engine

Welcome to Google App Engine

Before getting started, you want to learn more about developing and deploying applications.
Learn more about Google App Engine by reading the Getting Started Guide, the FAQ, or the Developer's Guide.

Create Application

Figure 7-6. Click on the 'Create Application' button to create your new Application hosted by the Google App Engine

By doing so, you will be then requested to verify your account by SMS (given that you have never used Google App Engine with your Google account in the past). After completing this required step, you will be forwarded to the main interface for creating an Application as illustrated in Figure 7-7.

Figure 7-7. The web interface for setting up a new Application in Google App Engine

On this page you need to enter an identifier for your application (it is like a URL name which will be also used for accessing your application). It can be anything you like and might represent your own project. Make sure you check the id you provide is available and also add a title for your new Application. You do not have to enter or modify any other information, so once done click on 'Create Application' button.

That's it! Your application is already registered on the Google App engine. The page you see now (see Figure 7-8) is the main web interface for managing your new Application.

Figure 7-8. The main web interface for managing your application on the Google App Engine

From this page you can manage things like application and user permissions, billing information, view the datastore indexes, view statistics about the usage of your application and much more.

For this project you do not need to make any changes, so you can move directly on to configuring Eclipse and writing the application code.

Setup Eclipse for Developing your Google App Engine Application
Although you can develop, compile, test and upload your App Engine application manually (using Java command line tools) the most convenient way for is to use the Eclipse Java Development Environment. Google provides the appropriate plugin for the latter, the Google Plugin for Eclipse. The plugin includes everything you need to build, test and deploy your app, entirely within the Eclipse environment.

The plugin is available for all the latter Eclipse versions (3.3 - 3.6). You can install the plugin using the Software Update feature of Eclipse. The installation locations are as follows:

- The Google Plugin for Eclipse, for Eclipse 3.3 (Europa): http://dl.google.com/eclipse/plugin/3.3
- The Google Plugin for Eclipse, for Eclipse 3.4 (Ganymede): http://dl.google.com/eclipse/plugin/3.4
- The Google Plugin for Eclipse, for Eclipse 3.5 (Galileo): http://dl.google.com/eclipse/plugin/3.5
- The Google Plugin for Eclipse, for Eclipse 3.6 (Helios): http://dl.google.com/eclipse/plugin/3.6

To get more help in installing the Google Plugin for your Eclipse environment visit: http://code.google.com/appengine/docs/java/tools/eclipse.html

After the setup, restart you Eclipse to activate the plugin and make it available and move on with creating a new App Engine Project.

Select the File menu > New > Web Application Project. The "Create a Web Application Project" wizard will open. For "Project name," enter a name for your project, such as 'GoogleAppsSensor' and for "Package," enter an appropriate package name, such as 'yourlastname.googleapps'. Then uncheck "Use Google Web Toolkit" and make sure that "Use Google App Engine" is checked (see Figure 7-9) and click 'Finish'.

Figure 7-9. The Create a Web Application Project Wizard Window

The wizard creates a directory structure for the project, including a 'src' directory for Java source files, and a 'war' directory for compiled classes and other files for the application, libraries, configuration files, images and CSS, and other data files. The wizard also creates a Servlet source file after the name of you project (like GoogleAppsSensorServlet) and two configuration files.

The Main Java Servlet

The Servlet (GoogleAppsSensorServlet.java) is the main Servlet that will handle the requests from your Arduino. Double click on it so that the source file opens in the Eclipse editor and add the code you will find for this project in the book's online code repository. Due to size restrictions, the code is not included in the chapter but you can find the complete project in the repository. The code explanation follows.

The Main Java Servlet – Code Review

The code starts by importing the appropriate libraries for handling communication, getting the current date, implementing the servlet and with the Google essential libraries for managing data storage.

```
import java.io.IOException;
import java.util.Date;
import javax.servlet.http.*;
import com.google.appengine.api.datastore.DatastoreService;
import
com.google.appengine.api.datastore.DatastoreServiceFactory;
import com.google.appengine.api.datastore.Entity;
import com.google.appengine.api.datastore.Key;
import com.google.appengine.api.datastore.KeyFactory;
```

Then you define the main Servlet class that extends the HttpServlet.

```
public class GoogleAppsSensorServlet extends HttpServlet
```

Since you will be using a GET request for receiving data from the Arduino you need to implement the doGet() function. The function takes two input variables, an HttpServletRequest object and an HttpServletResponse object. The server automatically invokes the function when a client request arrives. The HttpServletRequest object contains information about the GET request as received from the client and the HttpServletRequest object can be used to generate data sent back to the client. In this case you are not generating any data back to the Arduino, so you will not be using this object in your code.

```
public void doGet(HttpServletRequest req,
HttpServletResponse resp)throws IOException
```

However the HttpServletRequest object (req variable) contains all the information you need to manage from the Servlet (i.e. the sensor name and its current value). The latter information is passed as parameters into the GET request, so you read and store these parameters by invoking the getParameter() method as follows:

```
String SensorName = req.getParameter("sensor");
```

```
String value = req.getParameter("value");
```

You also create a date instance object that will be used in order to store the date and time of the incoming sensor value into the Datastore.

```
Date date = new Date();
```

You then implement the appropriate code for storing the sensor values to your application's provided datastore by the App Engine. To do so you need first to create a key for referring to the data (something like a database name). You use the createkey() method of the KeyFactory class. The first parameter of the method is a name for the key (like a database name) and the second is the ID of the latter. For the id you use the name of the sensor name that is provided by the GET request. With this implementation it is as if you would create a different database instance for every new sensor name. You do not need to implement any checking for existing key IDs when a new request arrives with an existing sensor name.

```
Key SensorDataKey =
KeyFactory.createKey("SensorData", SensorName);
```

You then need to create an Entity for storing the data (like a table in a relationship database). You give a name the Entity, e.g., "Sensor". Using the Entity object you define two the properties that will store the sensor data: one for the current date and time and one for the sensor value.

```
Entity sensorData = new Entity("Sensor",
    SensorDataKey);
sensorData.setProperty("date", date);
sensorData.setProperty("value", value);
```

Finally, you create a DatastoreService object using the DatastoreServiceFactory and you store the created Entity.

```
DatastoreService datastore =
DatastoreServiceFactory.getDatastoreService();
datastore.put(sensorData);
```

The servlet is now ready to accept sensor values through GET requests from the Arduino and store them into the Engine Datastore. Before uploading and deploying the Servlet you need to create the visualization JSP and edit a couple of files more.

The Visualization Java Server Page
As you have already mentioned, you need a JSP page to visualize the stored sensor values on the Cloud. The JSP page will be part of the Servlet deployment package, the 'war' folder. Right click on the latter (in the

Eclipse Package Explorer under your project's main structure) and select 'New File'.

Name the file something like 'sensorgraph.jsp' and click Finish. The JSP source file is now opened in the editor. Add the code of the file 'sensorgraph.jsp' that you will find in the source code folder for this chapter.

In order to visualize the data you make use of the Google Chart API, which is a JavaScript-based API for creating graphs. All the essential code for generating the graph is included into the JSP file. You can also define a specific CSS style for the generated web page. To do so you need to create a CSS file. Right-click again on the 'war' folder and select 'New->Folder'. Name it after 'stylesheets'. This new folder will also appear in the structure of your project under the 'war' folder. Now right-click on the 'stylesheets' folder and create a new file named 'main.css' and add the following code.

```
body {
    font-family: Verdana, Helvetica, sans-serif;
    background-color: #FFFFCC;
}
```

When done, the structure of your project (containing all the essential files for your App Engine application) shall look like the structure in Figure 7-10.

Figure 7-10. The structure of my Web Application Project when creating my App Engine Application

The Visualization Java Server Page – Code Review
A JSP file mainly consists of HTML code, but also includes Java code in special segments annotated with tags like '<%@' and '%>'. The annotations are used so that the Web container knows where the Java code

that needs to be compiled before the page is generate to the client, begins and ends respectively.

You begin the code by telling the Web container to add some information about the web page and its content. More specifically, you define the page content as text/html, the language encoding as UTF-8 and that the code between the special tags is Java code.

```
<%@ page contentType="text/html;charset=UTF-8"
language="java" %>
```

You then proceed with importing all essential Java libraries that will be used in the rest of the code (as you would do with coding a standard Java application). Most of them are related to the Google App Engine and the DataStore Service (since the main Java-based functionality of the Servlet will be to query the DataStore for sensor values).

```
<%@ page import="java.util.List" %>
<%@ page
import="com.google.appengine.api.datastore.DatastoreServiceF
actory" %>
<%@ page
import="com.google.appengine.api.datastore.DatastoreService"
%>
<%@ page import="com.google.appengine.api.datastore.Query"
%>
<%@ page import="com.google.appengine.api.datastore.Entity"
%>
<%@ page
import="com.google.appengine.api.datastore.FetchOptions" %>
<%@ page import="com.google.appengine.api.datastore.Key" %>
<%@ page
import="com.google.appengine.api.datastore.KeyFactory" %>
```

Then you start adding the HTML code that will visualize the sensor readings. Inside the HEAD segment of the HTML code you need to implement the JavaScript code that will utilize the Google Chart API for creating the sensor value graph.

```
<html>
  <head>
    <link type="text/css" rel="stylesheet"
href="/stylesheets/main.css" />
```

To better understand this part of the JSP code let's see first an example of how the Google Chart JavaScript code must look into the HTML file in order the graph to be properly illustrated:

Listing 7-10. A Google Chart Javascrip Code Sample

```
<script type="text/javascript"
src="https://www.google.com/jsapi"></script>
<script type="text/javascript">
```

```
  google.load("visualization", "1",
{packages:["corechart"]});
  google.setOnLoadCallback(drawChart);
  function drawChart() {
    var data = new google.visualization.DataTable();
    data.addColumn('string', 'Time');
    data.addColumn('number', 'Light');

    data.addRows(2);

    data.setValue(0, 0, 'Sun Nov 13 17:55:19 UTC 2011');
    data.setValue(0, 1, 685);

    data.setValue(1, 0, 'Sun Nov 13 17:55:12 UTC 2011');
    data.setValue(1, 1, 680);

    var chart = new
    google.visualization.LineChart(document.getElementBy
    Id('chart_div'));
    chart.draw(data, {width: 600, height: 240, title: '
    My Arduino Light Sensor Readings'});
  }
</script>
```

The Javascript code starts by importing the Chart API library (https://www.google.com/jsapi) and defining some essential API parameters. Then you implement the drawChart() function which is invoked when the page is loaded. It defines a DataTable variable (named data) that has two properties: a string property named Time and a number property named after the sensor name (Light in this case). The DataTable variable will contain all the sensor data (date and value) to be visualized. This example contains two sensor readings only, therefore you define two columns for the size of the DataTable variable. This as you might guess, is dynamically set by JSP code based on the DataStore saved data on the App Engine.

You then set the values of the DataTable variable using the setValue() method that takes 3 parameters. The first one is the incremental number of the value to graph starting from 0 (you have only two values so it is 0 and 1). The second one is the number of the property and the third the property value. In this case you have two properties that are data is plotted according to: the date and the sensor value.

Then a chart object is created defining parameters like the graph type (LineChart in this case) and the HTML element where the graph will be visualized (a div element named 'chart_div' in this example). Finally, the graph is visualized using the draw() method and by defining parameters like the width and height of the graph and the graph title.

So, based on this explanation, inside the drawChart() method you need first to define dynamically the Sensor name for the second attribute (the numeric value) of the DataTable variable.

You use the Java code tags and get the 'sensor' parameter from the client's GET request as follows:

```
data.addColumn('number',
'<%=request.getParameter("sensor")%>');
```

Next you need to define the number of rows. So the appropriate Java code must be inserted that will query the DataStore and determine the available number of sensor readings. You define a counter variable for that purpose and also request again the name of the sensor as parameter part of the GET request.

```
<%
int counter=0;
String SensorName = request.getParameter("sensor");
if (SensorName == null) {
    SensorName = "default";
}
```

You create a DataStoreService object and a Key object as you did in the Servlet code. Then you make a query for the "Sensor" Entity requesting the 10 most recent data entries sorted by the 'date' field.

```
DatastoreService datastore =
DatastoreServiceFactory.getDatastoreService();
Key SensorDataKey =
KeyFactory.createKey("SensorData", SensorName);

Query query = new Query("Sensor",
SensorDataKey).addSort("date",
Query.SortDirection.DESCENDING);
```

You create a List object with the results of the query. You also add the number of rows in the addRows Javascript method by using the size() method of the List object.

```
List<Entity> readings =
datastore.prepare(query).asList(FetchOptions.Builder.
withLimit(10));
%>
data.addRows(<%=readings.size()%>);
<%
```

Next comes the usage of the Javascript setValue() method in order to set the graph data. Using the counter variable, for every single object in the list you retrieve the 'date' and 'value' properties and insert them properly in the HTML code:

```
for (Entity reading : readings) {
%>
  data.setValue(<%=counter%>, 0, '<%=
  reading.getProperty("date") %>');
  data.setValue(<%=counter%>, 1, <%=
  reading.getProperty("value") %>);
  <%
  counter++;
}
```

You define the chart variable and initialize it properly and also define the title of the graph based on the sensor name.

```
var chart = new
google.visualization.LineChart(document.getElementById('
chart_div'));
chart.draw(data, {width: 600, height: 240, title: 'My
Arduino <%=SensorName%> Sensor Readings'});
```

Finally, inside the HTML body you add the div that will display the graph:

```
<body>
<p></p>
<div id="chart_div"></div>
</body>
</html>
```

Now that you have completed the Application code you are ready to deploy it on the Google App Engine!

Deploy your Application on the App Engine
To deploy your Application to the Google App Engine, you need first to set the application name to the appengine-web.xml file (under WEB-INF folder) as in the following example:

```
<application>arduinosensorcloud</application>
<version>1</version>
```

As you may notice, this is the ID you defined for your Application when creating the latter in the Google App Engine environment.

You also need to define in your Application what will be the URL names that will correspond to the servlet and the JSP file you just created. To do so you need to edit the web.xml file under the WEB-INF/lib folder of your project's structure. Assuming that the servlet for sending sensor values will be provided by the URL: http://yourapplicationname.appspot.com/add and the graph JSP page by: http://yourapplicationname.appspot.com/sensrograph you need to edit the web.xml similar to what follows:

Listing 7-11. A Google Chart Javascrip Code Sample

```xml
<?xml version="1.0" encoding="utf-8"?>
<web-app xmlns:xsi="http://www.w3.org/2001/XMLSchema-
instance"
xmlns="http://java.sun.com/xml/ns/javaee"
xmlns:web="http://java.sun.com/xml/ns/javaee/web-
app_2_5.xsd"
xsi:schemaLocation="http://java.sun.com/xml/ns/javaee
http://java.sun.com/xml/ns/javaee/web-app_2_5.xsd"
version="2.5">
        <servlet>
                <servlet-name>GoogleAppsSensor</servlet-
                name>
                <servlet-
                class>doukas.googleapps.GoogleAppsSensorServ
                let</servlet-class>
        </servlet>
        <servlet-mapping>
                <servlet-name>GoogleAppsSensor</servlet-
                name>
                <url-pattern>/googleappssensor</url-pattern>
        </servlet-mapping>
        <welcome-file-list>
                <welcome-file>sensorgraph.jsp</welcome-file>
        </welcome-file-list>

        <servlet>
        <servlet-name>add</servlet-name>
        <servlet-
        class>doukas.googleapps.GoogleAppsSensorServlet</ser
        vlet-class>
    </servlet>
    <servlet-mapping>
        <servlet-name>add</servlet-name>
        <url-pattern>/add</url-pattern>
    </servlet-mapping>
</web-app>
```

The final step for deploying your Application is the easiest one. After installing the Google App Plugin for Eclipse you may have notice a small blue icon with a 'g' letter on the left of your Eclipse toolbar. This is the 'Google Services and Development Tools' shortcut menu. Click on that and several options will appear as in Figure 7-11.

Figure 7-11. Using the 'Google Services and Development Tools' menu to
directly deploy your Application on the App Engine

Select 'Deploy to App Engine…' and click 'Deploy' on the next screen.
The first time you perform this you will be asked with your Google account
credentials. Then the Application will be automatically compiled, archived
into a 'war' file, uploaded and deployed on the App Engine.

Once the process completes successfully you will see a 'Deployment
completed successfully' message in your Eclipse console window.

That's it!

To test your deployed application (before your Arduino) open your
browser and enter the following URL:

```
http://arduinosensorcloud.appspot.com/add?sensor=test&value=
680
```

Remember to replace the 'arduinosensorcloud' with the name of your
Application. Re-enter the URL 3-4 times more and each time change the
sensor value from 680 to something different. Then call the sensorgraph
JSP page providing the 'test' as name sensor to see the visualization result:

```
http://arduinosensorcloud.appspot.com/sensorgraph.jsp?sensor
=test
```

It should look something like the output in Figure 7-12.

For more information on how you can utilize the Google App Engine
and improve your Cloud-based projects you can advise the books like
"Beginning Java Google App Engine" by Kyle Roche and Jeff Douglas.

Send Sensor Readings using the Ethernet Shield

You have successfully developed, uploaded, deployed and tested your
Application for visualizing sensor data on Google App Engine. Let's move
on to the last step of this project, communicating your Arduino directly to
the Application.

In order to provide the essential Internet connectivity to the Arduino,
you will use an Ethernet Shield or an Ethernet board. The following sketch

will demonstrate how to make the GET request to your Application on the App Engine, using the Arduino Ethernet Library. With the appropriate modifications you can create a sketch based on WiFi Library that does the same.

Let's assume that you have an analog sensor (e.g., a photoresistor or a temperature analog sensor) on the 0 analog pin of your Arduino board. You can also check the examples of Chapter 2 or the projects in the following examples and use a digital sensor instead.

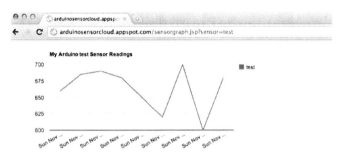

Figure 7-12. The Visualized Graph with the test sensor values

The Arduino Sketch

As in previous examples, upload the following code to your Ethernet-enabled Arduino and connect it with an Ethernet cable to an Internet accessible router. Then visit the sensorgraph JSP page on your App Engine URL and watch the sensor readings be visualized.

Listing 7-12. The Arduino Sketch Using The Ethernet Library for Sending Sensor Readings to Your App Engine Application

```
#include <SPI.h>
#include <Ethernet.h>

byte mac[] = { 0xDE, 0xAD, 0xBE, 0xEF, 0xFE, 0xED };
byte ip[] = { 192, 168, 1, 200 }; // a valid IP on your LAN

char server[] = "appspot.com";
EthernetClient client;

int pin = 0; // analog pin
int sensor = 0; // temperature variable

void setup()
{
```

```
    Ethernet.begin(mac, ip);
}
void loop()
{

    //Read sensor values
    sensor = analogRead(pin);
    String s = String(sensor,DEC);
    if (client.connect(server, 80)) {
      client.println("GET /add?value=" + s + "&sensor=Light
      HTTP/1.1");
      client.println("Host:arduinosensorcloud.appspot.com");
      //here is your app engine url - app id with appspot.com
      client.println();
      client.stop();
    }
    //send data every 10 seconds
    delay(10000);
}
```

The Arduino Sketch – Code Review

The beginning of the sketch is the same as in all previous examples that utilize the Arduino Ethernet Library. Notice that you use the 'appspot.com' as the name of the server in this case.

```
char server[] = "appspot.com";
EthernetClient client;
```

You also use two variables for the sensor readings, one defining the analog port (port 0 of the Arduino in this case) and one for storing the readings from the analog port.

```
int pin = 0; // analog pin
int sensor = 0; // temperature variable
```

As usually, the setup() function initializes the Ethernet module by binding the defined MAC with the IP address in the network.

The loop() function reads the analog value from the sensor and convers it into a String variable (using the String constructor).

```
//Read sensor values
sensor = analogRead(pin);
String s = String(sensor,DEC);
```

When the client is connected on the appspot.com on port 80 it will start making an appropriate HTTP GET request on the server. After defining the GET request, you define the path of the page you request, which in the case of the application server is '/add'. Then you add the parameters, which are the sensor name (e.g., Light in this example) and the sensor value stored in the string variable.

```
if (client.connect(server, 80)) {
   client.println("GET /add?value=" + s + "&sensor=Light
   HTTP/1.1");
```

You also need to specify to the Google Appspot web server that the exact application you are looking for is the name of your App Engine application. This is defined in the 'Host:' parameter sent over the web server.

```
client.println("Host:arduinosensorcloud.appspot.com");
```

You then send an empty line to the server to declare the end of the request and you also terminate the connection (since you are not expecting any data back from the server).

```
client.println();
client.stop();
```

Finally, you add a small delay of 10 seconds in the end of the loop so that you do not continuously send data over the Network:

```
delay(10000);
```

Summary

This Chapter has introduced you to Internet communication with your Arduino. After reviewing briefly some of the basic concepts for Internet communication, you were presented with examples of how the Arduino can be 'Internetized' using a wired Ethernet-based network, a wireless network and other means like communicating through your computer.

You have been also guided on how to develop your own Cloud Application that stores sensor readings and visualizes them, utilizing the Google App Engine.

Time to explore some of the most popular Cloud-based services for managing your Arduino sensor readings online!

PART III

USING THE ARDUINO TO MANAGE SENSOR DATA ON THE CLOUD

"The Cloud is for Everyone. The Cloud is a Democracy"

Marc Benioff, CEO - Salesforce.com

8 INTRODUCING COSM AS A CLOUD SERVICE

As discussed in Chapter 3, Cosm (former Pachube) has been one of the first available Cloud-based services for managing sensor data. It provides a light API for sending data directly from sensors and its web environment allows the visualization of data in graphs.

This chapter will take us further into exploring the Cosm service by learning the basic concepts and steps needed to store your sensor data. We will discuss about data concepts (feeds and datastreams), and triggers. We then will describe how to upload temperature, humidity and light data using your Arduino and different communication modules; an Ethernet shield, a WiFi-enabled Arduino and using your Android phone by exploiting ADB (Android Debug Protocol) or a Bluetooth connection. In addition, you will be presented with examples on how to manage a web-based trigger by Cosm, we explore the various data visualization options offered by Cosm and finally, an Android app is presented for monitoring your Cosm data feeds on the go.

Introduction to basic Cosm concepts

Before starting to put sensors together and writing some Arduino code, you shall take a brief introduction to the basic concepts of Cosm and familiarize with terms you will meet when storing and managing your data on this cloud-based online repository.

Cosm receives streams of sensor data organized in feeds, datapoints and datastreams.

- Feed (or Environment): A Cosm 'feed' is the data representation of an environment (e.g., your room) and its datastreams (e.g., your room's temperature and light conditions). Cosm allows you to use some metadata for optionally specifying whether it is physical or virtual, fixed or mobile, indoor or outdoor, etc.

- Datastream: A datastream represents a sensor (or generally any measuring device) within an environment. Every datastream must have a unique (within the environment) alphanumeric ID. It can also specify 'units' (e.g. 'watts') as well as user-defined 'tags' (e.g. 'fridge_energy').

- Datapoint: A datapoint represents a single value of a datastream at a specific point in time. It is simply a key value pair of a timestamp and the value at that time.

Consider this example: You want to monitor your room ('environment') with temperature, humidity and light sensors ('datastreams'). You create a Cosm feed titled 'My Room, with three datastreams where the IDs could be: 'temperature', 'humidity' and 'light'; which might be tagged 'thermal, non-contact', 'capacitive, DHT22' and 'LDR' respectively; and have units 'Celsius', '%RH' and 'Lux'. Individual datapoints at a point in time might be '28.0', '45' and 800 respectively.

Pull or Push Data

The Cosm Cloud service provides two ways for retrieving sensor data. Devices – sensors can either push (post) data to the service using its web service and data formatted either in JSON, XML or CSV format, or the service itself can pull (retrieve) data from the sensors. For the first option you need to implement Arduino code that will connect to the Cosm service, format the data to CSV and send them via POST protocol. On the second case you need to implement a Web Server on the Arduino with public access from the Internet enabled, so that the Cosm service can reach the Arduino through a URL you provide and retrieve the data.

Pull or Push? Depends on your needs and your implementation environment. Push is easier to implement and does not interfere with network firewall rules (since access to WWW port 80 is commonly allowed) but also requires a timing mechanism for updating the feed regularly. On the other hand, with Pull Cosm will acquire data from your Arduino on a predefined frequency, but you need to make sure your Arduino-based web server is accessible from the Internet.

Triggers

Triggers are used whenever you wish trigger an event based on your sensor readings. They are sort of thresholds you define and tell the Cosm service that when a specific datastream reaches, exceeds or changes from the threshold it shall invoke a URL. Cosm will send a sending HTTP POST request to that URL so that you can invoke a Web Service of yours and send some data as well (e.g., a flag telling your service to activate something).

According to Cosm, by default the minimum interval between sending a trigger notification twice is 5 seconds. Triggers fire only once when the condition is met and do not repeat in case the condition remains in this threshold.

In this Chapter you explore the use of triggers by using them for activating a relay switch based on sensor readings.

Data formats and Structures

The Cosm API currently allows you to communicate with it using 3 different data formats; JSON, XML and CSV. Which one you will use depends on your application needs. In the v2 API the JSON and XML representations of an environment both represent the data (your sensor readings) and metadata (e.g., Feed description, location information, etc.) This means that you can update the latter directly from your application using JSON or XML messages. CSV is used only for updating datastream values.

JSON

The JavaScript Object Notation, or simply JSON is a text-based open standard for data exchange, designed to be both computer and human-readable. It is derived from the JavaScript scripting language and is used for representing simple data structures called objects.

The JSON data format is considered very suitable for data exchange between web based applications as it can be easily parsed in user's browser by JavaScript. Another great benefit of JSON is that it produces much lower processing overheads than XML and uses less bandwidth to transmit, making it suitable for often data updates. Looking over a JSON message (like in Listing 8-1) you will notice two kinds of structures used: a) a collection of name/value pairs (e.g., "status": "live"), b) an ordered list of values (e.g., "max_value":"68.4", "min_value":"0.0","current_value":"27.79").

An example representation of a JSON-formatted message related to Cosm service is listed here:

Listing 8-1. An example of a Cosm feed in JSON-formated message.

```
{
"website":"https://sites.google.com/site/mycloudsensor/",
"status":"live",
"location":
        {"disposition":"fixed",
        "domain":"physical",
        "lat":38.0014721525229,"
        exposure":"indoor",
        "lon":23.8211059570312},
"description":
        "My Cloud-based collection of sensor data and
        management of indoor temperature, Humidity, light.
        Information retrieved from DHT22 temperature and
        humidity sensor and a custom photoresistor and sent
        throught a BlackWidow (WiFi enabled arduino)",
"feed":
        "https://api.cosm.com/v2/feeds/28602.json",
```

```
    "creator":"https://cosm.com/users/harisdmac",
    "datastreams":
        [{"tags":["Temp"],"unit":{"type":"basicSI","la
bel":"Celcius","symbol":"C"},"max_value":"32.9","at":
"2011-09-
18T14:15:52.854615Z","min_value":"0.0","current_value
":"30.29","id":"0"},
        {"tags":["Humidity"],"unit":{"type":"basicSI",
"label":"%","symbol":"%"},"max_value":"68.4","at":"20
11-09-
18T14:15:52.854615Z","min_value":"0.0","current_value
":"27.79","id":"1"},
        {"tags":["Light"],"unit":{"type":"basicSI","la
bel":"0-
1024","symbol":"U"},"max_value":"1023.0","at":"2011-
09-
18T14:15:52.854615Z","min_value":"0.0","current_value
":"836","id":"2"}]],
"private":"false",
"updated":"2011-09-
18T14:15:52.854615Z","id":28602,"title":"MyCloudSensor","ver
sion":"1.0.0"

}
```

XML

XML does not need any particular introduction. The Cosm API uses a specific format of XML called EEML (See http://www.eeml.org for more details). An example of a full XML representation looks like the following:

Listing 8-2. A full XML representation of a Cosm fee using EEML format

```
<eeml xmlns="http://www.eeml.org/xsd/0.5.1" xmlns:xsi="http:
//www.w3.org/2001/XMLSchema-
instance" version="0.5.1"xsi:schemaLocation="http://www.eeml
.org/xsd/0.5.1 http://www.eeml.org/xsd/0.5.1/0.5.1.xsd">
        <environment updated="2011-09-
        18T14:20:50.788234Z" id="28602" creator="https://cosm
        .com/users/harisdmac">
                <title>MyCloudSensor</title>
                <feed>https://api.cosm.com/v2/feeds/28602.x
                ml</feed>
                <status>live</status>
                <description>
                        My Cloud-based collection of sensor data
                        and management of: Indoor temperature
                        Indoor Humidity Indoor light Information
                        retrieved from DHT22 temperature and
                        humidity sensor and a custom
                        photoresistor and sent throught a
                        BlackWidow (WiFi enabled arduino)
                </description>
                <website>https://sites.google.com/site/mycloud
```

```
sensor/</website>
<private>false</private>
<location domain="physical" exposure="indoor"
disposition="fixed">
        <lat>38.0014721525229</lat>
        <lon>23.8211059570312</lon>
</location>
<data id="0">
        <tag>Temp</tag>
        <current_value at="2011-09-
18T14:20:50.788234Z">30.29</current_valu
e>
        <max_value>32.9</max_value>
        <min_value>0.0</min_value>
        <unit type="basicSI" symbol="C">Celcius<
/unit>
</data>
<data id="1">
        <tag>Humidity</tag>
        <current_value at="2011-09-
18T14:20:50.788234Z">25.60</current_valu
e>
        <max_value>68.4</max_value>
        <min_value>0.0</min_value>
        <unit type="basicSI" symbol="%">%</unit>
</data>
<data id="2">
        <tag>Light</tag>
        <current_value at="2011-09-
18T14:20:50.788234Z">838</current_value>
        <max_value>1023.0</max_value>
        <min_value>0.0</min_value>
        <unit type="basicSI" symbol="U">0-
1024</unit>
</data>
    </environment>
</eeml>
```

Same information contained in the JSON message is included in the XML formatted message.

CSV

CSV is a comma-separated data format mostly used for textual representations. It is very suitable for use by embedded devices, such as an Arduino or other low powered microcontroller. It contains none of the metadata that the XML and JSON formats contain (though this can be added separately using the API or the web interface). A full representation of CSV is as follows:

Listing 8-3. CSV values for a Cosm feed

```
0,2011-09-18T14:45:50.484032Z,30.29
1,2011-09-18T14:45:50.484032Z,25.20
2,2011-09-18T14:45:50.484032Z,817
<feed_id>,<datastream_id>,<timestamp>,<value>
```

The default format in Cosm for handling feeds is JSON. These different data formats are used in cases where users want to retrieve Cosm feed data and parse them on their own applications, or send data about a feed from their own application to the Cosm service. CSV looks to be the most simple formatting but as compared to XML and JSON, but it lacks of additional information (like feed - sensor location, etc.). As you will see in our project samples, CSV will be used for directly communicating Arduino with Cosm, whereas for more advanced communication, external libraries can be utilized for helping us with the data formatting.

Private and Public Data Access

Cosm allows the creation of private or publicly available feeds. Public means that through the Cosm web site, all users can view your feeds. To make sure no one tampers your sensor readings, a feed can only be maintained (updated, modified and deleted) by the owner of the feed. To do so they have introduced the usage of API keys. The keys are categorized into the Master API key and Secure Sharing Keys. On the contrary, Private feeds are only visible to you.

The Master API Key provides full access to all features currently available on users account, including editing, updating, retrieving, creating and deleting Cosm feeds. A Master API Key should not be kept private. It is provided to users upon registration and is available at the My API keys page.

The Secure Sharing key(s) can be used in case users wish to share a key with specific restrictions on the key that controls the way it can be used (consequently and the access rights to the feed data).

Create and manage feeds

So the first thing one needs to do for storing sensor data on Cosm is to create a feed. Cosm offers a simple web interface for creating feeds, by adding additional information, datastreams and creating triggers and managing them. Alternatively you can use the API to create and manage feeds remotely using your own application.

For example, you can create the following feed in JSON format:

Listing 8-4. A new feed in JSON format.

```json
{
  "title":"My Feed",
  "website":"http://www.mysite.com/",
  "version":"1.0.0",
  "tags":[
      "Tag1",
      "Tag2"
  ],
  "location":{
    "disposition":"fixed",
    "ele":"23.0",
    "name":"room",
    "lat":38.00147,
    "exposure":"indoor",
    "lon":23.82110,
    "domain":"physical"
  },
  "datastreams":[
    {
      "current_value":"28.3",
      "max_value":"32.0",
      "min_value":"10.0",
      "tags":["temperature"],
      "id":"0"
    },
    {
      "current_value":"987",
      "max_value":"90.0",
      "min_value":"10.0",
      "tags":["humidity"],
      "id":"1"
    }
  ]
}
```

and save it into a text file naming it after 'myfeed.txt'. Then you can use the curl command line application (available in Linux/Unix systems and MacOSX, windows users can find an executable here: http://curl.haxx.se/download.html) to setup the new feed in Cosm:

```
curl --request POST --data-binary @myfeed.txt --header "X-
CosmApiKey: YOUR_API_KEY_HERE" http://api.cosm.com/v2/feeds
```

Notice that you need your Master API Key. Upon execution of the command application, the URL of the newly created feed will be returned using the "Location" header, or more simply you can just browse your Cosm account and find your new feed there!

Cosm offer two types of feeds; manual and automatic.

Manual vs Automatic Feeds

Manual feeds are those where your sensor device 'pushes' updates, usually via a PUT Web request. Usually you perform this by defining your own time intervals, or based on value-changes (or events like a button press). Manual Feeds are more useful for projects that sit behind a firewall or that are too low-powered to host a Web server (you need a Web server for automatic feeds, since as explained below in such case Cosm service requests data from the device).

Automatic feeds, on the other hand, are those where the environment or device is able to serve data on request: Cosm automatically 'pulls' data from them either every 15 minutes or whenever another client requests it (whichever is the more frequent).

The status of a feed can be either live or frozen, which can mean slightly different things depending on the feed type:

- Manual:
  - Live means that the remote environment or device manually updated Cosm (i.e. 'pushed') with its data in the last 15 minutes.
  - Frozen means that it was last updated more than 15 minutes ago.
- Automatic:
  - Live means that Cosm has successfully retrieved data (i.e. 'pulled') from the remote environment or device in the last 15 minutes.
  - Frozen means that either it has not retrieved data in the last 15 minutes or that the last time it attempted to retrieve data (within the last 15 minutes) it was unsuccessful.

Let's move on with your first project for uploading feed data on Cosm!

Project 1 - Use Arduino to upload feed data from environmental sensors

Your first project exploring the Cosm data service will be about storing indoor temperature, humidity and light conditions as acquired by your favorite Arduino. The data collected will be transmitted to a Cosm feed you will create for visualizing the data and retrieving some data history. You will split the project into three basic steps; the first step will require the construction of the circuit, the second step concerns the feed setup at the Cosm environment and finally, the third step will include the Arduino code needed for collecting and transmitting the sensor data.

As described in previous chapters, there many ways to connect Arduino with the Internet and thus with Cosm and you mainly focused on using an Ethernet shield (equivalent to an Ethernet-enabled Arduino), a WiFi shield and using an Android phone acting as an intermediate node that collects information from Arduino and forwards the latter to the Internet. You will be presented with code samples for all three alternatives!

What you will need for the basic circuit:

1. An Arduino board or a clone (board can be Uno, Mega or compatible)
2. A DHT22 temperature and humidity sensor
3. A LDR photoresistor
4. A 10K Ohm resistor
5. A 47K Ohm resistor
6. A power source for the Arduino (can be your USB port or an external power source like a 9V battery)

For the connection to the Internet: You can use either a WiFi shield (or a WiFi-enabled Arduino), an Ethernet shield or an ADK board and an Android phone.

For easy assemble and avoiding soldering the parts you can use:

- A breadboard
- Jumper wires

The parts above are enough for building a circuit that will measure temperature, humidity and light using the sensors.

Project 1 - The circuit setup

The DHT22 sensor is a digital sensor and the LDR photoresistance is analog, therefore you will need a digital and an analog port from the

Arduino respectively. Instead of connecting the sensor outputs directly to the ports, you need to set up a small circuit first.

The LDR acts as a resistor and has two pins (no polarity issues in resistors). You connect one pin (let's name it pin A) directly to the 5V power output of the Arduino and the either pin (pin B) through the 10K Ohm resistor to the Ground (GND) port of the Arduino. You then connect a jumper from a point between pin B and the resistor to the analog port 0 of the Arduino. This way you can read from the analog port any voltage dropouts caused by the LDR depending on light presence.

The reason to put the resistor in between is to make sure you get valid readings even if the LDR is removed or indicates limited resistance.

The DHT22 sensor comes 4 pins but only 3 of them make use to us. The two of them are for creating a circuit and circulating current, therefore are connected to 5V and Ground pins of the Arduino respectively. The third pin is the sensor output pin that will generate information about the sensed temperature and humidity. Figure 8-1 provides more details on the pins and their usage. You place a 4.7K Ohm resistor between the power (5V) pin and the output pin for the same reason you did with the LDR. You then connect the output pin directly to the digital port 7 of your Arduino. You can check Figure 8-2 for a complete breadboard setup visualization and Figure 8-3 for the detailed schematic of the circuit that illustrates connections more clearly.

Figure 8-1. The DHT22 humidity and temperature sensor. Pin 1: Vdd power supply, Pin 2: Data signal, Pin 3: Not used, Pin 4: Ground connection.

You are ready! Powering the Arduino will also power the circuit and sensors will start to deliver information to the Arduino.

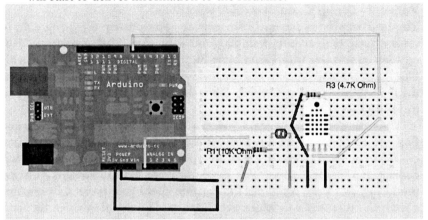

Figure 8-2. Breadboard design for the Project 1 circuit

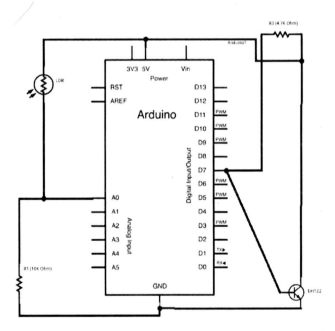

Figure 8-3. The actual schematic for the circuit. You can notice the connections between the resistors and the LDR and the DHT22 sensor, as well the inputs in the Arduino.

Project 1 - Set up Cosm to receive your data

The next step is to set up Cosm service appropriately for being able to receive your temperature, humidity and light data. This involves initially creating an account with Cosm here: http://www.cosm.com. Click the big blue 'Get Started' button and provide the required information (email, desired username and password). Since the end of 2011 Cosm offers all of its services for free.

Upon successful registration you can visit the 'My API Keys' page and find out your Master API Key that you will use in order to upload feed data.

Next you need to create a feed for your environment sensors. Feed will contain information about the three different sensor streams you are producing, and some additional information about your project. By creating a feed, Cosm provides us with a very important information about the feed, the Feed ID. Using the Feed ID you can always refer to your feed and retrieve information about your sensors, but most importantly you need it for storing the information from the sensors in the first place.

To create a new Cosm feed, click on the button. You will see four options: a) Arduino, b) Current Cost NetSmart, c) Twitter Stats and d) Something Else.

While it is easier to use the first option for setting up an Arduino feed, I suggest to try first the general option (Something Else). A new dialog like the one in Figure 8-4 will ask you about feed details, like whether Cosm shall pull data or not, Feed title and Feed tags.

Figure 8-4. Setting up the Cosm feed. We enter general data like title, description, location, etc.

On the last step click the 'Create' button. You feed is now created and a dialog similar to the one below will inform you about your Feed's ID.

Figure 8-5. Creating your new Feed!

Click 'Finish'. Now you need to create a Datastream for your Feed. Click the [+ Datastream] button. The new page asks for Datastream information like the ID, Tags, Units and Symbol for Units.

As explained before, datastreams represent to data acquired by sensors. So, each datastream corresponds to a specific sensor. Therefore you need to create three different datastreams as illustrated in Figure 8-6. It is not necessary to add all the displayed information about the datastream (tags, units and symbol) but at least an ID is required for each one.

When finished, just click the "Save Changes" button. That's it! You have just created your first feed in Cosm service. The next page showing up will contain the important Feed ID, you can find it either on the URL of the page (it will be something similar to https://cosm.com/feeds/28602) and under the 'WEBSITE' tag. The feed is up and can receive and store datastream data from your sensors! Next you will see the code essentials for making that happen.

Datastreams

ID	Tags	Units	Symbol	
0	Temp	Celsius	c	✖ Remove
1	Humidity	%	H	✖ Remove
2	Light	lux	L	✖ Remove

Figure 8-6. Defining the temperature, humidity and light datastreams for your feed.

Project 1 - Connect to the Web Service and send data

You have set up your circuit, created a feed and the appropriate datastreams in Cosm. The remaining step is to program your Arduino for sending the data over Cosm so you can watch them being visualized in graphs and more.

As already discussed, you will cover three communication ways: a) using an Ethernet shield (identical to Ethernet-enabled Arduino), b) using a WiFi shield (or a WiFi-enabled Arduino) and c) using an Android phone. For the latter case, you will also demonstrate two different ways of sending data to the phone: 1) With a direct USB connection using the Android Debug Bridge (ADB) as an alternative to Google's ADK and an ADK-enabled Arduino board, 2) Using a Bluetooth connection.

Using the Arduino Ethernet or the Ethernet Shield

To use a wired Ethernet communication with Cosm you can utilize an Ethernet Arduino board (http://arduino.cc/en/Main/ArduinoBoardEthernet) or a shield like one in Figure 8-7. The rest of the circuit remains the same as described in previous section.

When selecting an Ethernet shield for your Arduino, make sure it is based on the 'Wiznet Ethernet' chip supported by the official Arduino Ethernet library.

Figure 8-7. A Seeedstudio Ethernet shield based on the Wiznet W5100 Ethernet chip (image courtesy of Seeedstudio).

Make sure you connect the Ethernet port with an Ethernet cable to a LAN network using a switch or a router. Then upload the following code to your Arduino board. You can find the source code for this project at **http://www.buildinginternetofthings.com**.

Listing 8-5. Code for Project 1 using an Arduino Ethernet or the Ethernet Shield.

```
#include <DHT22.h>
#include <SPI.h>
#include <Ethernet.h>

// Data wire is plugged into port 7 on the Arduino
#define DHT22_PIN 7
```

```
// Setup a DHT22 instance
DHT22 myDHT22(DHT22_PIN);
DHT22_ERROR_t errorCode;

//Useful char buffers for creating the CSV datastream
content
char buf1[16];
char buf2[16];
char DataBuff[16];

//Light sensor is connected to analog port 0
#define LIGHT_PIN        0

// Time variable for custom timer
long lastTime;

// sensor variables
float temp;
float humid;
int light;

//Set up ethernet and network related data
byte mac[] = { 0xCC, 0xAC, 0xBE, 0xEF, 0xFE, 0x91 };
byte ip[] = { 192, 168, 1, 124 }; // no DHCP so we set our
own IP address
byte CosmServer[] = { 173, 203, 98, 29 };

//create a Client object for handling connection
Client localClient(CosmServer, 80);

//feed data function, provides data for the POST command in
comma separated values (CSV)
void feedData()
{
   ftoa(buf1, temp, 2); //we need to convert float temp to
   //char
   ftoa(buf2, humid, 2);    //we need to. convert float humid
   //to char
   sprintf(DataBuff,"%s,%s,%d", buf1, buf2,light); //we save
   //all variables including light into one char variable
   //DataBuff

}

//Define the sendData() function that handles the
communication with Cosm implementing the protocol for
sending feeds based on Cosm API v2.
void sendData(){
  if (localClient.connect()) {
    feedData();
    int content_length = strlen(DataBuff);
```

```
    localClient.print("PUT /v2/feeds/");
    localClient.print("28602");
    localClient.print(".csv HTTP/1.1\nHost:
    api.pachube.com\nX-PachubeApiKey: ");
    localClient.print("NkyX---90s"); //Replace with your
                                     Cosm API Key
    localClient.print("\nUser-Agent: ");
    localClient.print("\nContent-Type: text/csv\nContent-
    Length: ");
    localClient.print(content_length);
    localClient.print("\nConnection: close\n\n");
    localClient.print(DataBuff);
    localClient.print("\n");
  }
}


//Arduino setup function
void setup(void)
{

  //initialize sensor variables
  temp = 0.0f;
  humid = 0.0f;
  light = 0;

  pinMode(LIGHT_PIN, INPUT);

  //initiate Ethernet module
  Ethernet.begin(mac, ip);
  delay(500);

  //Note start time
  lastTime = millis();
}

//Arduino loop function
void loop(void)
{

    if ((millis() - lastTime) > 20000){
       //timer has expired
       lastTime = millis();

       //Read DHT22 Data here:
       errorCode = myDHT22.readData();

       switch(errorCode){
       case DHT_ERROR_NONE:
           temp =  myDHT22.getTemperatureC();
           humid = myDHT22.getHumidity();
       break;
```

```
    }

    light = analogRead(LIGHT_PIN);

    //send the data over Cosm
    sendData();
  }

   //small delay for the main loop
   delay(100);
}

//Convert double to char (due to currently sprintf in
Arduino fails to do so)
char *ftoa(char *a, double f, int precision)
{
  long p[] =
{0,10,100,1000,10000,100000,1000000,10000000,100000000};

  char *ret = a;
  long heiltal = (long)f;
  itoa(heiltal, a, 10);
  while (*a != '\0') a++;
  *a++ = '.';
  long desimal = abs((long)((f - heiltal) * p[precision]));
  itoa(desimal, a, 10);
  return ret;
}
```

Using the Arduino Ethernet or the Ethernet Shield – Code Overview

```
#include <DHT22.h>
```

The DHT22 sensor has a digital output. This means you have to interpret the incoming bits to bytes and then to useful information related to sensed temperature and humidity. To save yourself from the effort and the additional coding you will use an open library available for the Arduino provided by Ben Adams (https://github.com/nethoncho/Arduino-DHT22). To install the library, just download the files from the source repository and place them under ../Arduino installation folder/Resources/Java/Libraries in a folder named DHT22.

There is also an alternative library for the DHTxx family sensors, with similar installation and usage found here: http://arduino.cc/playground/Main/DHTLib

```
#include <SPI.h>
#include <Ethernet.h>
```

The Arduino team provides us with an official Ethernet library (http://arduino.cc/en/Reference/Ethernet) that simplifies the communication with the Internet (i.e. connecting to ports, communication protocols, etc.) with simple functions. Using the Ethernet library requires also importing the SPI library.

Next, you define several variables that deal with the DHT22 sensor, a time variable used for a custom timer and sensor variables about the temperature, humidity and light:

```
// Data wire is plugged into port 7 on the Arduino
#define DHT22_PIN 7

// Setup a DHT22 instance
DHT22 myDHT22(DHT22_PIN);
DHT22_ERROR_t errorCode;

//Useful char buffers for creating the CSV datastream
content
char buf1[16];
char buf2[16];
char DataBuff[16];

//Light sensor is connected to analog port 0
#define LIGHT_PIN        0

// Time variable for custom timer
long lastTime;

// sensor variables
float temp;
float humid;
int light;
```

Next you set up Ethernet variables for implementing communication with the Cosm service:

```
//Set up ethernet and network related data
byte mac[] = { 0xCC, 0xAC, 0xBE, 0xEF, 0xFE, 0x91 };
byte ip[] = { 192, 168, 1, 124 }; // We need to set your own
IP address, make sure this address can be connected to your
local gateway/router
byte CosmServer[] = { 173, 203, 98, 29 }; //the IP address
of Cosm service
```

You shall notice the local IP that the Ethernet module will use to connect to the Internet through your local gateway/router. Make sure you change it to whatever is necessary. You will probably also notice that you do not use URLs like www.cosm.com but IP addresses instead. This is because the Ethernet module and the library do not support DNS resolving yet (conversion of URL to IP address).

The following line in code creates a Client object (like a client socket) that will connect to the Cosm service based on the IP you have provided and port 80 (default WWW port).

```
//create a Client object for handling connection
Client localClient(CosmServer, 80);
```

You are done with the variables that will be needed by your entire program so you now declare essential functions that will be used. You start with the feedData() function that construct the CSV message containing the sensor values and store it into the DataBuff variable.

```
//feed data function, provides data for the POST command in
comma separated values (CSV)
void feedData()
{
    ftoa(buf1, temp, 2);
    ftoa(buf2, humid, 2);
    sprintf(DataBuff,"%s,%s,%d", buf1, buf2,light);
}
```

You continue with the sendData() function that implements all the communication that needs to be done with the Cosm service:

```
//Define the sendData() function that handles the
communication with Cosm implementing the protocol for
sending feeds based on Cosm API v2.
void sendData(){
  if (localClient.connect()) {
    feedData();
    int content_length = strlen(DataBuff);
    localClient.print("PUT /v2/feeds/");
    localClient.print("28602");
    localClient.print(".csv HTTP/1.1\nHost:
    api.pachube.com\nX-PachubeApiKey: ");
    localClient.print("NkyX---90s"); //Replace with your
                                        Cosm API Key
    localClient.print("\nUser-Agent: ");
    localClient.print("\nContent-Type: text/csv\nContent-
    Length: ");
    localClient.print(content_length);
    localClient.print("\nConnection: close\n\n");
    localClient.print(DataBuff);
    localClient.print("\n");
  }
}
```

If you have a successful connection (localClient.connect()) you then invoke the feedData() function that will construct the CSV message containing the sensor values and store it into the DataBuff variable.

Then you say to the Cosm web server you have connected to, that we want to PUT (POST) data to the address /v2/feeds/28602.csv which obviously is your Feed ID. By .csv you declare to the service that we will be using CSV as a data format. You then add some header data and include your Cosm API key.

You also define the content type of the post data (text/csv), the length of the data you are posting (that exist in the DataBuff variable) and finally send some additional information about the http connection.

That's all you need to communicate with the Cosm service!

The following lines contain the standard Arduino setup and loop functions.

```
//Arduino setup function
void setup(void)
{

    //initialize sensor variables
    temp = 0.0f;
    humid = 0.0f;
    light = 0;

    pinMode(LIGHT_PIN, INPUT);

    //initiate Ethernet module
    Ethernet.begin(mac, ip);
    delay(500);

    //Note start time
    lastTime = millis();
}
```

Inside the setup function you initialize your sensor variable with null values, you set the 0 analog port (LIGHT_PIN) to INPUT so that Arduino knows you are expecting some analog data from there.

You also initialize the Ethernet module by binding it to the mac and IP address you have provided. You allow a short delay there to make sure the module is initialized properly before you move on.

Finally, you note the starting time of your program so that you can later check inside the loop() function if the desired time has elapsed and you can send data to Cosm (like a timer).

```
//Arduino loop function
void loop(void)
{

    //checks if 20 seconds have ellapsed
    if ((millis() - lastTime) > 20000) {
        // timer has expired
        lastTime = millis();
```

```
   //Read DHT22 Data here:
   errorCode = myDHT22.readData();

   switch(errorCode){
   case DHT_ERROR_NONE:
       temp = myDHT22.getTemperatureC();
       humid = myDHT22.getHumidity();
   break;
 }

 light = analogRead(LIGHT_PIN);

 //send the data over Cosm
 sendData();
}

//small delay for the main loop
delay(100);
}
```

The loop() function starts with checking if the timer has expired (set to 20000 msec) so you can continue with acquiring and uploading the sensor data. If so, you set the timer (lastTime) to current time for next usage, and you proceed with reading data from the DHT22 sensor through the library you are using. You also read the light sensor data through the analog port (LIGHT_PIN).

Once done, all data are saved into the appropriate variables (temp, humid and light) and you proceed with invoking the sendData() function for transmitting the latter to the Cosm service. Outside the if statement of the timer check, you add a small delay of 100msec to the main Arduino loop.

Using a WiFi shield or a WiFi-enabled Arduino
The next part of the project is to examine how to wirelessly update the Cosm feed using a WiFi shield or a WiFi-enabled Arduino such as the BlackWidow from AsyncLabs (see Figure 8-8). The pin connections on the Arduino are the same as described in the circuit setup section.

When using the WiFi shield or the BlackWidow board make sure it is in the range of a wireless router capable of providing access to the Internet.

Figure 8-8. Breadboard with LDR and DHT22 sensors connected to a WiFi-enabled Arduino.

Keep also in mind that you can find the complete Arduino source code online at **http://www.buildinginternetofthings.com**

Listing 8-6. Code for sending feeds to Cosm using a WiFi Arduino Shield.
```
/*
* A simple sketch that uses WiFi to PUT (via POST) to Cosm
*/
#include <DHT22.h>
#include <WiServer.h>
#include <MsTimer2.h>

// Data wire is plugged into port 7 on the Arduino
#define DHT22_PIN 7

// Setup a DHT22 instance
DHT22 myDHT22(DHT22_PIN);
DHT22_ERROR_t errorCode;

//Wireless configuration defines --------------------------
------------
#define WIRELESS_MODE_INFRA    1
```

```
#define WIRELESS_MODE_ADHOC    2

// Wireless configuration parameters ------------------
unsigned char local_ip[] = {192,168,1,200};    // IP address
of WiShield
unsigned char gateway_ip[] = {192,168,1,1};    // router or
gateway IP address
unsigned char subnet_mask[] = {255,255,255,0};    // subnet
mask for the local network
const prog_char ssid[] PROGMEM = {"YOUR ACCESS POINT NAME"};
// max 32 bytes

unsigned char security_type = 3;    // 0 - open; 1 - WEP; 2 -
WPA; 3 - WPA2

// WPA/WPA2 passphrase
const prog_char security_passphrase[] PROGMEM = {"XXXX"};
// max 64 characters
// WEP 128-bit keys
prog_uchar wep_keys[] PROGMEM = {
    0x00, 0x00, 0x00, 0x00, 0x00, 0x00, 0x00, 0x00, 0x00,
    0x00, 0x00, 0x00, 0x00,    // Key 0
    0x00, 0x00, 0x00, 0x00, 0x00, 0x00, 0x00, 0x00, 0x00,
    0x00, 0x00, 0x00, 0x00,    // Key 1
    0x00, 0x00, 0x00, 0x00, 0x00, 0x00, 0x00, 0x00, 0x00,
    0x00, 0x00, 0x00, 0x00,    // Key 2
    0x00, 0x00, 0x00, 0x00, 0x00, 0x00, 0x00, 0x00, 0x00,
    0x00, 0x00, 0x00, 0x00}; // Key 3
// setup the wireless mode
// infrastructure - connect to AP
// adhoc - connect to another WiFi device
unsigned char wireless_mode = WIRELESS_MODE_INFRA;
unsigned char ssid_len;
unsigned char security_passphrase_len;
// End of wireless configuration parameters -------------

//non WiShield defines
char buf1[16];
char buf2[16];
char buf3[16];

//light sensor is attached to analog port 0
#define LIGHT_PIN        0

//global state data
boolean intTimer;

// sensor variables
float temp;
float humid;
int light;
```

```
// IP Address for Cosm.com
uint8 ip[] = {173,203,98,29};
char hostName[] = "www.cosm.com\nX-PachubeApiKey: NkyX-kR7O-
--------------------------AS90s\nConnection: close";
char url[] = "/api/feeds/28602.csv?_method=put";
// A request that POSTS data to Cosm
POSTrequest postCosm(ip, 80, hostName, url, feedData);


//function for initializing the timer
void timerISR() {
    intTimer = true;
}

//body data function, provides data for the POST command in
comma separated values (CSV)
//currently POSTs one value but more can be added by
separating with comma (no spaces)
void feedData()
{

    ftoa(buf1, temp, 2);
    ftoa(buf2, humid, 2);
    sprintf(buf3,"%s,%s,%d", buf1, buf2,light);
    WiServer.print(buf3);
}

void setup(void)
{

 //Initiliaze the sensor variables
  temp = 0.0f;
  humid = 0.0f;
  light = 0;

    //setup the analog pins as input
    pinMode(LIGHT_PIN, INPUT);

    // Initialize WiServer (we'll pass NULL for the page
serving function since we don't need to serve web pages)
    WiServer.init(NULL);

    //set up global state data
    intTimer = false;

    //make one initial measurement and send it:
    DHT22_ERROR_t errorCode;
    errorCode = myDHT22.readData();

    switch(errorCode)
    {
      case DHT_ERROR_NONE:
```

```
      temp =  myDHT22.getTemperatureC();
      humid = myDHT22.getHumidity();
      break;
  }
  light = analogRead(LIGHT_PIN);
  postCosm.submit();
  WiServer.server_task();

   //setup the timerISR to be called every minute
   MsTimer2::set(300000, timerISR); // 5min period
   MsTimer2::start();
}

void loop(void)
{

   //handle the timer expire
   if(true == intTimer) {
      // timer has expired

      //set it to false again so timer restarts
      intTimer = false;

      // Read DHT22 sensor data:
      errorCode = myDHT22.readData();

      switch(errorCode)
  {
    case DHT_ERROR_NONE:
       temp =  myDHT22.getTemperatureC();
       humid = myDHT22.getHumidity();
       break;
        }

   light = analogRead(LIGHT_PIN);

      //submit data
      postCosm.submit();

   }
   //need to make sure WiServer is active and completes its
tasks
    WiServer.server_task();

   //add a small delay for the loop
   delay(100);
}

//Convert double to char (due to currently sprintf in
Arduino fails to do so)
char *ftoa(char *a, double f, int precision)
{
```

```
long p[] =
{0,10,100,1000,10000,100000,1000000,10000000,100000000};

char *ret = a;
long heiltal = (long)f;
itoa(heiltal, a, 10);
while (*a != '\0') a++;
*a++ = '.';
long desimal = abs((long)((f - heiltal) * p[precision])));
itoa(desimal, a, 10);
return ret;
}
```

Using a WiFi shield or a WiFi-enabled Arduino – Code Overview

```
#include <DHT22.h>
#include <WiServer.h>
#include <MsTimer2.h>
```

Same as in previous code, you start with importing the DHT22 library for parsing the sensor data. Next you have the WiFi server library that will allow us to communicate easily over HTTP protocol with the Cosm service. The library can be obtained from here https://github.com/asynclabs/WiShield and is installed the common way external Arduino libraries are.

The next library imported (MsTimer2.h) is an easy to use timer. It is used here as an alternative to counting elapsed time using millis() (like in Listing 8-5). You just need to set the desired interval time in ms, and declare the function that will be invoked each time the timer expires. In this case, you set a variable to true, that is later checked by the main execution loop for uploading feed data. The library can be obtained from here: http://www.arduino.cc/playground/Main/MsTimer2

As with previous code you setup the variables for parsing the DHT22 data:

```
// Data wire is plugged into port 7 on the Arduino
#define DHT22_PIN 7

// Setup a DHT22 instance
DHT22 myDHT22(DHT22_PIN);
DHT22_ERROR_t errorCode;
```

Then you define the variables and parameters related to the WiFi shield for the WiFi communication:

```
//Wireless configuration defines ------------------------
#define WIRELESS_MODE_INFRA    1
#define WIRELESS_MODE_ADHOC    2

// Wireless configuration parameters ---------------------
```

```
unsigned char local_ip[] = {192,168,1,200};   // IP address
of WiShield
unsigned char gateway_ip[] = {192,168,1,1};   // router or
gateway IP address
unsigned char subnet_mask[] = {255,255,255,0};   // subnet
mask for the local network
const prog_char ssid[] PROGMEM = {"YOUR ACCESS POINT NAME"};
// max 32 bytes

unsigned char security_type = 3;   // 0 - open; 1 - WEP; 2 -
WPA; 3 - WPA2

// WPA/WPA2 passphrase
const prog_char security_passphrase[] PROGMEM = {"XXXX"};
// max 64 characters
// WEP 128-bit keys
prog_uchar wep_keys[] PROGMEM = {
    0x00, 0x00, 0x00, 0x00, 0x00, 0x00, 0x00, 0x00, 0x00,
    0x00, 0x00, 0x00, 0x00,   // Key 0
    0x00, 0x00, 0x00, 0x00, 0x00, 0x00, 0x00, 0x00, 0x00,
    0x00, 0x00, 0x00, 0x00,   // Key 1
    0x00, 0x00, 0x00, 0x00, 0x00, 0x00, 0x00, 0x00, 0x00,
    0x00, 0x00, 0x00, 0x00,   // Key 2
    0x00, 0x00, 0x00, 0x00, 0x00, 0x00, 0x00, 0x00, 0x00,
    0x00, 0x00, 0x00, 0x00}; // Key 3
// setup the wireless mode
// infrastructure - connect to AP
// adhoc - connect to another WiFi device
unsigned char wireless_mode = WIRELESS_MODE_INFRA;
unsigned char ssid_len;
unsigned char security_passphrase_len;
// End of wireless configuration parameters ---------------
```

Make sure you enter the correct name of your Access Point,
you select the proper security type of your WiFi connection
(WEP, WPA or WPA2) and you enter the correct security
passphrase at the appropriate variables.
Next you add some variables for reading the sensor values
and composing the CSV data:

```
//non WiShield defines
char buf1[16];
char buf2[16];
char buf3[16];

//light sensor is attached to analog port 0
#define LIGHT_PIN        0

//global state data
boolean intTimer;

// sensor variables
float temp;
float humid;
```

```
int light;
```

You then add and initialize the essential variables for making the communication with Cosm service:

```
// IP Address for Cosm.com
uint8 ip[] = {173,203,98,29};
char hostName[] = "www.cosm.com\nX-PachubeApiKey: NkyX-kR7O-
----------------------------AS90s\nConnection: close";
char url[] = "/api/feeds/28602.csv?_method=put";
// A request that POSTS data to Cosm
POSTrequest postCosm(ip, 80, hostName, url, feedData);
```

Make sure you place your Master API key properly in the hostName variable! Also, notice that the URL variable contains the Feed ID as well. You define the method type as 'put', since you want to push data to Cosm and therefore you also create a POSTrequest object.

Next you move on with the essential and additional functions for your main Arduino program. You start with a function invoked when the MsTimer2 expires and sets the intTimer variable to true, so that the main execution loop is aware of when to upload feed data:

```
//function for initializing the timer
void timerISR() {
    intTimer = true;
}
```

The following function (feedData) creates the CSV that contains the sensor values in the proper format and stores it into the global variable buf3. It also prints the CSV to the network stream through the WiServer.print() command.

```
//body data function, provides data for the POST command in
comma separated values (CSV)
//currently POSTs one value but more can be added by
separating with comma (no spaces)
void feedData()
{

    ftoa(buf1, temp, 2);
    ftoa(buf2, humid, 2);
    sprintf(buf3,"%s,%s,%d", buf1, buf2,light);
    WiServer.print(buf3);
}
```

Next in the code follows the usual Arduino setup command where you initialize variables and the WiFi module:

```
void setup(void)
{
```

```
//Initiliaze the sensor variables
  temp = 0.0f;
  humid = 0.0f;
  light = 0;
```

You set up the analog pin 0 as input:

```
pinMode(LIGHT_PIN, INPUT);
```

and you initialize WiServer (you'll pass NULL for the page serving function since you don't need to serve web pages):

```
WiServer.init(NULL);
```

You set up global state data:

```
intTimer = false;
```

and make one initial measurement and send it:

```
DHT22_ERROR_t errorCode;
errorCode = myDHT22.readData();

switch(errorCode)
{
  case DHT_ERROR_NONE:
    temp = myDHT22.getTemperatureC();
    humid = myDHT22.getHumidity();
    break;
}
light = analogRead(LIGHT_PIN);
postCosm.submit();
WiServer.server_task();
```

You finally setup the timerISR to be called every 5 minutes

```
MsTimer2::set(300000, timerISR); // 5min period
MsTimer2::start();
```

Next follows the Arduino loop function that will check if the 5 minute timer has expired, will read sensor values and will forward them to the Cosm service through the appropriate function calls.

```
void loop(void)
{

    //handle the timer expire
    if(true == intTimer) {
    // timer has expired
    //set it to false again so timer restarts
    intTimer = false;
```

```
//READ DHT22 Data here:
 Serial.print("Requesting data...");
 errorCode = myDHT22.readData();

 switch(errorCode){
    case DHT_ERROR_NONE:
        temp = myDHT22.getTemperatureC();
    humid = myDHT22.getHumidity();
    break;
    }

 light = analogRead(LIGHT_PIN);

 //submit data
 postCosm.submit();


 }
 //need to make sure WiServer is active and completes its
 //tasks
 WiServer.server_task();

 //add a small delay for the loop
 delay(100);
}
```

Finally, you implement the help function for converting double values (for temperature and humidity sensor readings) to char:

```
//Convert double to char (due to currently sprintf in
Arduino fails to do so)
char *ftoa(char *a, double f, int precision)
{
  long p[] =
{0,10,100,1000,10000,100000,1000000,10000000,100000000};

  char *ret = a;
  long heiltal = (long)f;
  itoa(heiltal, a, 10);
  while (*a != '\0') a++;
  *a++ = '.';
  long desimal = abs((long)((f - heiltal) * p[precision]));
  itoa(desimal, a, 10);
  return ret;
}
```

Using your Android phone and Android Debug Bridge (ADB)

This section describes how to use your Android phone and the ADB to forward sensor data to Cosm service. In order to have your Android phone communicate with Arduino you need a compatible ADK board, like the official Arduino ADK board (http://arduino.cc/en/Main/ArduinoBoardADK) or the Seeeduino ADK Main Board.

The project will use exactly the same circuit setup regarding the sensors. For this case you will need code for the Arduino to communicate with the Android phone through ADB and an appropriate Android application that will receive the sensor information and update the Cosm feed respectively. In addition, the Android application visualizes the sensor readings (temperature, relative humidity and light) and also provides a button for checking connection with the ADK board (see Figure 8-9). The button controls the Arduino on-board led (pin 13) and gives us also a nice chance to demonstrate communication from the Android towards the Arduino board.

The following listing contains the Arduino sketch for communicating and sending sensor data to the phone.

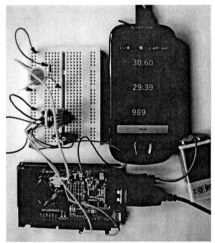

Figure 8-9. Photo of my setup for communication between Android ADK board and an Android phone. The board is Seeeduino ADK (from seeedstudio.com) and requires external powering (9V battery attached as shown). The Android app visualizes the data from the sensors and also updates the Cosm feed via WiFi connection.

Sending sensor data to Android – Arduino Code

Listing 8-7. Code for sending sensor reading to Android phone using the ADK

```
#include <SPI.h>
#include <Adb.h>
#include <DHT22.h>

// Adb connection.
Connection * connection;
```

```
// Elapsed time for ADC sampling. The rate at which ADC
value is sent to Android device.
long lastTime;

// Data wire is plugged into port 7 on the Arduino
#define DHT22_PIN 7

// Setup a DHT22 instance
DHT22 myDHT22(DHT22_PIN);
DHT22_ERROR_t errorCode;

#define LIGHT_PIN       0

//State of LED. Initially OFF.
uint8_t LEDState=0;

// sensor variables
float temp;
float humid;
int light;

char buf1[16];
char buf2[16];
char buf3[16];

// Event handler for the shell connection.
// This event handler is called whenever data is sent from
Android Device to Seeeduino ADK.
// Any data / command to be sent to I/O of ADK has to be
handled here.

void adbEventHandler(Connection * connection, adb_eventType
event, uint16_t length, uint8_t * data)
{

  // In this example Data packets contain one byte and it
  //decides the state of a LED connected to port 13
  // The size of data is predetermined for this application.
  Android device also uses the same size.

  if (event == ADB_CONNECTION_RECEIVE)
  {
    if(LEDState != data[0])
    {
        // Change the state of LED
        digitalWrite(13, data[0]);
        // Store the State of LED
        LEDState = data[0];
    }
  }
```

```
}

void setup()
{

  // Note start time
    lastTime = millis();

    temp = 0.0f;
    humid = 0.0f;
    light = 0;

    //setup the analog pins as input, probly redundant
    //pinMode(TEMPERATURE_PIN, INPUT);
    pinMode(LIGHT_PIN, INPUT);

    //Set Digital pin 13 (Arduino LED) as output
    pinMode(13,OUTPUT);

    //Initialise the ADB subsystem.
    ADB::init();

    //Open an ADB stream to the phone's shell. Auto-
    //reconnect. Use any unused port number eg:4568

    connection = ADB::addConnection("tcp:4568", true,
    adbEventHandler);

}

void loop()
{
    //make one initial measurement and send it:
    DHT22_ERROR_t errorCode;
    errorCode = myDHT22.readData();

    switch(errorCode)
    {
      case DHT_ERROR_NONE:
        temp =  myDHT22.getTemperatureC();
        humid = myDHT22.getHumidity();
        break;
    }
    light = analogRead(LIGHT_PIN);

    //Check if ADC needs to be sampled.
    if ((millis() - lastTime) > 20000)
    {
      ftoa(buf1, temp, 2);
      ftoa(buf2, humid, 2);
      sprintf(buf3,"%s,%s,%d", buf1, buf2,light);
```

```
    //Send the ADC value to Android device as two bytes of
    //data.
    connection->writeString(buf3);
    lastTime = millis();
  }

  // Poll the ADB subsystem.
  ADB::poll();
}

//Convert double to char (due to currently sprintf in
Arduino fails to do so)
char *ftoa(char *a, double f, int precision)
{
  long p[] =
{0,10,100,1000,10000,100000,1000000,10000000,100000000};

  char *ret = a;
  long heiltal = (long)f;
  itoa(heiltal, a, 10);
  while (*a != '\0') a++;
  *a++ = '.';
  long desimal = abs((long)((f - heiltal) * p[precision]));
  itoa(desimal, a, 10);
  return ret;
}
```

When compiling this code in Arduino environment make sure you select the appropriate board. The Arduino ADK and the Seeeduino ADK Main Board are Arduino Mega 2560 compatible.

Sending sensor data to Android – Arduino Code – Code Overview
Starting with the library imports, apart from the DHT22 library you need to include the essential ones for communicating with the Android phone.

```
#include <SPI.h>
#include <Adb.h>
#include <DHT22.h>
```

Google's ADK provides support for Android 2.3.4 and newer versions. In order to achieve communication with phones supporting older Android firmware (like 2.1) you will use MicroBridge instead. MicroBridge is an Android Debug Bridge (ADB) implementation for microcontrollers like the Arduino. The ADB protocol uses TCP sockets that enable establishing bidirectional pipes between the Arduino and the Android phone. More information about the ADB and the essential library files can be obtained from here: http://code.google.com/p/microbridge/

Next you create an Adb connection object that will be used to handle all the communication between the board and the phone:

```
Connection * connection;
```

You also define a timer for the Elapsed time for ADC sampling (the rate at which ADC value is sent to Android device):

```
long lastTime;
```

As usually, you need to tell Arduino the pins the sensors are plugged into:

```
#define DHT22_PIN 7

// Setup a DHT22 instance
DHT22 myDHT22(DHT22_PIN);
DHT22_ERROR_t errorCode;

#define LIGHT_PIN        0
```

You define the initial state of the LED as off.

```
uint8_t LEDState=0;
```

The last variables defined are as in previous code listing the variables for handling sensor reading and for creating the CSV message:

```
// sensor variables
float temp;
float humid;
int light;

char buf1[16];
char buf2[16];
char buf3[16];
```

The first function defined is the code is the event handler. The latter is called whenever data is sent from Android Device to Seeeduino ADK. Any data / command to be sent to I/O of ADK has to be handled here. You also implement the code that checks for if incoming data (data[0]) is 0 or 1 so that the state of the pin will be low or high:

```
void adbEventHandler(Connection * connection, adb_eventType
event, uint16_t length, uint8_t * data)
{

    // In this example Data packets contain one byte and it
       decides the state of a LED connected to port 13
    // The size of data is predetermined for this application.
       Android device also uses the same size.

    if (event == ADB_CONNECTION_RECEIVE)
```

```
{
    if(LEDState != data[0])
    {
        //Change the state of LED
        digitalWrite(13, data[0]);
        //Store the State of LED
        LEDState = data[0];
    }
}

}
```

Secondly, you implement the Android setup function that will initialize the timer (lastTime) with the current time (millis()) and also initializes the sensor variables.

```
void setup()
{

    // Note start time
    lastTime = millis();

    temp = 0.0f;
    humid = 0.0f;
    light = 0;

    //setup the analog pins as input, probly redundant
    // pinMode(TEMPERATURE_PIN, INPUT);
    pinMode(LIGHT_PIN, INPUT);
```

You set Digital pin 13 (Arduino LED) as output:

```
pinMode(13,OUTPUT);
```

You initialize the ADB subsystem:

```
ADB::init();
```

and you finally open an ADB stream to the phone's shell. You can use any unused port number e.g., 4568:

```
connection = ADB::addConnection("tcp:4568", true,
adbEventHandler);
```

The Arduino main loop will make the essential sensor readings, check if timer has expired, check if the button has been clicked (event received by the handler) and will forward data to the phone:

```
void loop()
{
    //make one initial measurement and send it:
    DHT22_ERROR_t errorCode;
    errorCode = myDHT22.readData();
```

```
  switch(errorCode)
  {
    case DHT_ERROR_NONE:
      temp = myDHT22.getTemperatureC();
      humid = myDHT22.getHumidity();
      break;
  }
  light = analogRead(LIGHT_PIN);

  //Check if ADC needs to be sampled.
  if ((millis() - lastTime) > 20000) {
    ftoa(buf1, temp, 2);
    ftoa(buf2, humid, 2);
    sprintf(buf3,"%s,%s,%d", buf1, buf2,light);

    //Send the ADC value to Android device as two bytes
      of data.
    connection->writeString(buf3);
    lastTime = millis();
  }

  //Poll the ADB subsystem.
  ADB::poll();
}
```

Next section describes the Android code for receiving and visualizing the latter data.

Sending sensor data to Android - Android code overview

Presenting and analyzing the complete Android code is out of the context of this book. The Listings and code overview include only the essential parts for enabling the communication between Arduino, the phone and the Cosm service. The complete Android project can be downloaded from http://www.buildinginternetofthings.com

You need to create an Android main activity that will handle the data exchange and also visualize data. You start by importing the essential libraries for ADB communication. In addition to that, we will also import an external library, JPachube, as an easy way to update a Cosm feed. JPachube is a Java wrapper for the Cosm API and can be obtained from here: http://code.google.com/p/jpachube/

```
import Pachube.Feed;
import Pachube.Pachube;
import org.microbridge.server.AbstractServerListener;
import org.microbridge.server.Server;
```

...

In the onCreate initialization method of Android we create the TCP server instance that will communicate with the board through USB.

```
@Override
public void onCreate(Bundle savedInstanceState) {
...
// Create TCP server (based on  MicroBridge LightYouight
//Server)
try{
server = new Server(4568); //Use the same port number
                        used in ADK Main Board firmware
  server.start();
} catch (IOException e){
  Log.e("Seeeduino ADK", "Unable to start TCP server", e);
}
```

We need to define a port for the server to start listening to. This TCP port will receive and send data to the board. We also call the start() server method that will initiate the server on the specified port. The server will be started on a separate thread immediately listening for connections.

We then add a listener to the sever so we can handle data when they are received from the board. For updating the Android UI and visualizing the sensor values we use a separate class (UpdateData):

```
server.addListener(new AbstractServerListener() {

@Override
public void onReceive(org.microbridge.server.Client
client, byte[] data){

    //We check the length of the data
    if (data.length<2) return;
    //and we create a String representation of the
    byte[] data received
    adcSensorValue = new String(data);

    runOnUiThread(new Runnable() {
        @Override
        public void run() {
            //We use a different AsyncTask class for
            //handling the UI updates (i.e. drawing the
            //sensor values)
            new UpdateData().execute(adcSensorValue);
        }
    });

}
```

```
});
...
}
```

You will utilize the JPachube API to update your feed. The data comes into CSV format from the board. You need to split it and use the updateDatastream method to update each datastream separately:

```
void Cosm(String data) {

try {

    String feeds[] = data.split(",");
    Pachube p = new Pachube("YOUR MASTER API KEY");
    Feed f = p.getFeed(28602);
    f.updateDatastream(0, Double.valueOf(feeds[0]));
    f.updateDatastream(1, Double.valueOf(feeds[1]));
    f.updateDatastream(2, Double.valueOf(feeds[2]));

    } catch (Exception e) {

    }
}

// UpdateData Asynchronously sends the value received from
ADK Main Board.
// This is triggered by onReceive()

class UpdateData extends AsyncTask<String, Integer, String>
{
// Called to initiate the background activity
    @Override
    protected String doInBackground(String... sensorValue) {
        return (String.valueOf(sensorValue[0]));
    }

    // Called once the background activity has completed
    @Override
    protected void onPostExecute(String result) {
        //Invoke the method for updating the feed:
        Cosm(adcSensorValue);
        String[] feeds = result.split(",");

        TextView tvAdcvalue = (TextView)
        findViewById(R.id.tvADCValue);
        tvAdcvalue.setText(String.valueOf(feeds[0]));

        TextView tvAdcvalue2 = (TextView)
        findViewById(R.id.tvADCValue2);
        tvAdcvalue2.setText(String.valueOf(feeds[1]));
```

```
    TextView tvAdcvalue3 = (TextView)
    findViewById(R.id.tvADCValue3);
    tvAdcvalue3.setText(String.valueOf(feeds[2]));

    }
}
```

Using your Android phone and Bluetooth Connection

The circuit setup in this case is again similar with the initial one (see Figure 8-2), you just need either a Bluetooth enabled Arduino or a Bluetooth shield. For making the communication between your Android phone and the Arduino board you can utilize the Amarino toolkit. The Amarino toolkit (http://www.amarino-toolkit.net/) provides libraries both for the Arduino and the Android that make the communication over Bluetooth easy and direct.

First you will need to pair your Bluetooth module with your mobile phone. Then you will need to import the Amarino library to your Arduino environment and upload the following sketch to your board:

Listing 8-8. Arduino Code for sending sensor reading to Android phone using a Bluetooth Connection.

```
#include <DHT22.h>
#include <MeetAndroid.h>

//create a MeetAndroid object for handling BT communication
MeetAndroid meetAndroid;

// Data wire is plugged into port 7 on the Arduino
#define DHT22_PIN 7

// Setup a DHT22 instance
DHT22 myDHT22(DHT22_PIN);
DHT22_ERROR_t errorCode;

#define LIGHT_PIN       0

char buf1[16];
char buf2[16];
char DataBuff[16];

// sensor variables
float temp;
float humid;
int light;

void setup()
{
  temp = 0.0f;
```

```
  humid = 0.0f;
  light = 0;

  //setup the analog pins as input
  pinMode(LIGHT_PIN, INPUT);
}

void loop()
{
  //make one initial measurement and send it:
   DHT22_ERROR_t errorCode;
   errorCode = myDHT22.readData();

   switch(errorCode)
   {
     case DHT_ERROR_NONE:
      temp =  myDHT22.getTemperatureC();
      humid = myDHT22.getHumidity();
      break;
   }
   light = analogRead(LIGHT_PIN);

   // send sensor measurements to Android as comma separated
   //values:
   ftoa(buf1, temp, 2);
   ftoa(buf2, humid, 2);
   sprintf(DataBuff,"%s,%s,%d", buf1, buf2,light);
   meetAndroid.send(DataBuff);


   // instead of using an advanced timer just delay the loop
   //for 10 seconds
   delay(10000);
}

//Convert double to char (due to currently sprintf in
Arduino fails to do so)
char *ftoa(char *a, double f, int precision)
{
  long p[] =
  {0,10,100,1000,10000,100000,1000000,10000000,100000000};

  char *ret = a;
  long heiltal = (long)f;
  itoa(heiltal, a, 10);
  while (*a != '\0') a++;
  *a++ = '.';
  long desimal = abs((long)((f - heiltal) * p[precision]));
  itoa(desimal, a, 10);
  return ret;
}
```

Depending on the Bluetooth module you are using, you might need to configure it to slave and inquiring mode first following the vendor's instructions.

Using your Android phone and Bluetooth Connection – Code Overview

In this case you will focus only on the important parts of the code that implement the Bluetooth communication using Amarino. You need to include the Amarino library (MeetAndroid.h) in the beginning of the code and before compiling the code to make sure the library is installed in Arduino environment.

```
#include <MeetAndroid.h>
```

You then create a MeetAndroid object that will handle all the communication with the phone:

MeetAndroid meetAndroid;

The rest of code declares the essential variables for reading sensor data and also initializes them within the setup() function. You do not need to do anything special regarding Amarino there.

You then use the send() method of the meetAndroid object for sending the sensor readings to the phone as comma separated values (using the same trick with DataBuff char array and the ftoa() method). That's it!

```
ftoa(buf1, temp, 2);
ftoa(buf2, humid, 2);
sprintf(DataBuff,"%s,%s,%d", buf1, buf2,light);
meetAndroid.send(DataBuff);
```

Using your Android phone and Bluetooth Connection – Android Code Overview

As with previous project, the full Android source is available online at **http://www.buildinginternetofthings.com**. Here you will see only the essential parts of the code handling the communication with the Arduino board through Bluetooth and the communication with Cosm.

You need to create an Android activity that will communicate with the Arduino board using the Amarino library. You start by importing the essential libraries. You also import JPachube library for handling directly the communication with Cosm:

```
...
import at.abraxas.amarino.Amarino;
import at.abraxas.amarino.AmarinoIntent;
import Pachube.Feed;
import Pachube.Pachube;
...
```

You need to set the MAC address of the Bluetooth module you are using:

```
private static final String DEVICE_ADDRESS =
"00:18:E4:0C:68:C5";
```

and you move on by defining the ArduinoReceiver object. This object contains all essential methods for communication over Bluetooth:

```
private ArduinoReceiver arduinoReceiver = new
ArduinoReceiver();
```

You also need to register this receiver in order to be able to receive broadcasted intents:

```
registerReceiver(arduinoReceiver, new
IntentFilter(AmarinoIntent.ACTION_RECEIVED));
```

The following line tells Amarino to connect to the specific Bluetooth device from within own code:

```
Amarino.connect(this, DEVICE_ADDRESS);
```

The following ArduinoReceiver class in the code is responsible for catching broadcasted Amarino events. It extracts data from the intent and updates the UI accordingly:

```
public class ArduinoReceiver extends BroadcastReceiver {
@Override
    public void onReceive(Context context, Intent intent) {
        String data = null;
        String[] parsed;

        //the device address from which the data was sent,
        //you don't need it here but to demonstrate how you
        //retrieve it
        final String address =
        intent.getStringExtra(AmarinoIntent.EXTRA_DEVICE_ADD
        RESS);

        // the type of data which is added to the intent
        final int dataType =
        intent.getIntExtra(AmarinoIntent.EXTRA_DATA_TYPE, -
        1);

        // you only expect String data though, but it is
        //better to check if really string was sent
        // later Amarino will support differnt data types,
        so //far data comes always as string and
        // you have to parse the data to the type you have
        //sent from Arduino, like it is shown below
        if (dataType == AmarinoIntent.STRING_EXTRA){
            data =
            intent.getStringExtra(AmarinoIntent.EXTRA_DATA);

            if (data != null){
                //Update feeds to Cosm:
```

```
        Cosm(data);

        //visualize them on phone:
        parsed = data.split(",");
        Sensor1.setText(parsed[0]);
        Sensor2.setText(parsed[1]);
        Sensor3.setText(parsed[2]);
    }
  }
}

//Update Cosm feed here using the JPachube library.
void Cosm(String data) {
    try {
        //Receive the data form board in CSV, so split
        it //to get the separate values
        String feeds[] = data.split(",");

        Pachube p = new Pachube("YOUR MASTER API KEY");
        Feed f = p.getFeed(28602);
        f.updateDatastream(0, Double.valueOf(feeds[0]));
        f.updateDatastream(1, Double.valueOf(feeds[1]));
        f.updateDatastream(2, Double.valueOf(feeds[2]));

        } catch (Exception e) {
            e.printStackTrace();
        }
    }
}
```

Project 2 - Control a relay switch using a Cosm trigger and your Arduino

Another important feature of the Cosm service is that it allows users to set triggers. A trigger can be defined as a command that will be executed when a user-specified rule is met. The rule has of course to do with the feeds you are managing through Cosm service. Triggers can be set for each separate datastream of a feed. Rules are defined by simple comparison of current datastream values with user-defined thresholds. When a rule is met, the trigger action is to invoke a URL defined by user.

An interesting application is the controlling remotely of a relay switch using a Cosm trigger and your Arduino. Consider this as an extension to the previous project that supplements its functionality with respect to the Internet of Things (IoT) notion. Remember in the initial chapters when talking about IoT devices and their functions; they do not only sense the environment but they are also able to interact with it. With this project you can extend the circuit to control an actuator (the relay switch) based on sensor readings. The sensor reading will be about the light condition in your home environment and the actuator will control the internal lights. The benefit of using a Cloud-based trigger is that you can modify the rule for the trigger anytime, anywhere without the need to modify either your circuit or your embedded code.

The circuit setup

As explained in Chapter 2, a relay is an electro-mechanical switch that opens and closes under the control of another electric circuit. When current flows through the coil of the relay, a magnetic field is created that causes an armature to move, either making or breaking an electrical connection. When current is removed from the relay coil, the armature returns to its rest position. It is important to place a diode across the coil of the relay because a spike of voltage is generated when the current is removed from the coil due to the collapse of the magnetic field. This voltage spike can damage the sensitive electronic components controlling the circuit.

To present code and circuit setup more clear, you will not interfere with the setup presented in Project 1. You will consider a separate Arduino circuit for controlling the relay that has also an Ethernet shield and Internet access.

Materials needed:

1. A relay switch
2. A 2N3904 NPN transistor
3. A 1N4001 diode

4.   An Ethernet Arduino or an Ethernet shield

Consider the schematic of Figure 8-10. The relay you are using in this example has a coil voltage of 5V (DC), meaning the relay switch will be activated when 5VDC is supplied across the relay coil. The output pin of an Arduino does supply 5V, but the current that the relay requires to activate its switch is greater than the Arduino can safely supply. In this case, a 2N3904 NPN transistor is used to supply a higher current 5V source to the relay coil. The digital port 7 of Arduino controls the NPN transistor. By 'high' state it will close the circuit with the Gnd and allow current to flow the 5V source towards the relay switch. The relay switch will then change state and close the upper circuit.

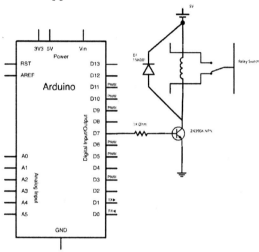

Figure 8-10. The circuit schematic for Project 2. The digital port 7 of Arduino controls the NPN transistor. By 'high' state it will close the circuit with the Gnd and allow current to flow the 5V source towards the relay switch. The relay switch will then change state and close the upper circuit.

This project involves two steps: First, you need to implement your own Web server on the Arduino using the Ethernet library, so that the Cosm trigger can communicate directly with it and order it to control the relay switch. The Web server will have a specific URL address than can be used by Cosm to reach it. The URL address will be similar to http://94.65.132.102/?relay=1 The IP is the address of the server that can be reached from the Internet. The 'relay' it is a token you will use in order to make sure you 'receive' a trigger - command from Cosm related to the switch. Final, the 1 is the Boolean code for switching the Arduino output to 'high' or 'low' state respectively.

Then, you need to create a trigger for one of your datastreams in Cosm and declare the aforementioned URL.

Create the Web Server on the Arduino Using Arduino Ethernet or the Ethernet Shield

Upload the following sketch to your Arduino. Make sure you enter an appropriate IP address that can be resolved from the Internet and that there is a connection with the Internet available.

Listing 8-9. Arduino Code for receiving triggers from Cosm service through the Ethernet library.

```
#include <SPI.h>
#include <Ethernet.h>
#include <TextFinder.h>

//relay switch pin
#define RELAY_PIN 7

//Web server essentials
#define maxLength 25 // preventing BoF
#define serverPort 80 // listening port
//Create the server instance
Server server(serverPort);

//Set up ethernet and network related data
byte mac[] = { 0xCC, 0xAC, 0xBE, 0xEF, 0xFE, 0x91 };
// We set our own IP address.
//Make sure the address can be resolved from the Internet
//using e.g., NAT or port forwarding or use a real IP
address instead
byte ip[] = { 94, 65, 132, 102 };

void setup(void)
{
  //Initialize the serial for debugging
  Serial.begin(9600);
  //Set pin to output for controlling the relay
  pinMode(RELAY_PIN, OUTPUT);

  //Initiate the Web Server
  Ethernet.begin(mac, ip);
  server.begin();
}

void loop(void)
{
  Client client = server.available();
  if (client) {
  TextFinder finder(client );
  while (client.connected()) {
    if (client.available()) {
```

```
// counters to show the number of pin change requests
int digitalRequests = 0;
int analogRequests = 0;
//We got a web page request from client:
if( finder.find("GET /") ) {
   // find tokens starting with "relay" and stop on the
   //first blank line
   while(finder.findUntil("relay", "\n\r")){
      //Get the value and check if it is 1 or 0 and
      //control the relay switch
      int val = finder.getValue();
      if(val==1) {
         Serial.println("On");
         digitalWrite(RELAY_PIN, HIGH);
      }
      else{
         Serial.println("Off");
         digitalWrite(RELAY_PIN, LOW);
      }
    }
   }
   //Just send something back to the client (Cosm
   //service in this case)
   client.println("HTTP/1.1 200 OK");
   client.println("Content-Type: text/html");

   client.println();

   //We've finished so we can exit the while loop and
   //disconnect client
   break;
  }
 }
 //Give browser some time
 delay(1);
 //Disconnect client
 client.stop();
 }
}
```

Code Overview
You start by importing the essential libraries for the Ethernet library and the Web server implementation:

```
#include <SPI.h>
#include <Ethernet.h>
#include <TextFinder.h>
```

You also include the TextFinder library. It is a very useful library for extracting information from a stream of data. It has been initially created for use with the Arduino Ethernet library to find particular fields and get strings or numeric values and it can be obtained from here: http://www.arduino.cc/playground/Code/TextFinder

You move on by defining the relay switch pin and the Web server essential variables (e.g., port 80 as listening, the server object, etc.):

```
#define RELAY_PIN 7

//Web server essentials
#define maxLength 25 // preventing BoF
#define serverPort 80 // listening port
String inString = String(maxLength);
Server server(serverPort);
```

You also set up Ethernet and network related data as in Project 1

```
byte mac[] = { 0xCC, 0xAC, 0xBE, 0xEF, 0xFE, 0x91 };
// You set our own IP address.
//Make sure the address can be resolved from the Internet
//using e.g., NAT or port forwarding or use a real IP
address instead
byte ip[] = { 94, 65, 132, 102 };
```

At the Arduino setup function you set the digital pin mode of the attached transistor pin to output and you also initialize the server instance:

```
pinMode(RELAY_PIN, OUTPUT);
//Initiate the Web Server
Ethernet.begin(mac, ip);
server.begin();
```

Now, the main work for the server that accepts clients and handles the incoming data is performed at the main execution loop. Initially you create a Client object for accepting incoming connections:

```
Client client = server.available();
```

If you have an incoming connection:

```
if (client) {
```

You use the TextFinder to search easily for tokens inside the client's web request:

```
TextFinder  finder(client);
```

And you parse the incoming data and search for the 'relay' token when you receive a 'GET' command for client (default request for web page):

```
while (client.connected()) {
   if (client.available()) {
      if( finder.find("GET /") ) {
```

You need to find tokens starting with "relay" and stop on the first blank line:

```
while(finder.findUntil("relay", "\n\r")){
```

You have found relay command, so control the relay switch based on the command value, 1 or 0:

```
int val = finder.getValue();
if(val==1) {
    Serial.println("On");
    digitalWrite(RELAY_PIN, HIGH);
}
else{
    Serial.println("Off");
    digitalWrite(RELAY_PIN, LOW);
}
```

Finally, you just send something back to the client (Cosm service in this case):

```
client.println("HTTP/1.1 200 OK");
client.println("Content-Type: text/html");
client.println();
```

At the end you also disconnect the client so that the (empty) web response is finished and served to the client:

```
client.stop();
```

Create a Cosm Trigger

To create a Cosm trigger you need to enter the feed main page, e.g. https://cosm.com/feeds/28602. Then you need to select a datastream for which you will create the trigger (let's say it is datastream 2). Move your mouse over datastream 2 so that the configuration options appear on the right. Click the 'Show Triggers' link. Beneath your datastream you will see a message saying 'There are no triggers yet for this datastream' and the '+Trigger' button next to it. Click the button. A new window will ask you for the type of trigger you wish to add: 'HTTP POST' or 'Twitter'. For this example select the HTTP POST option.

The window will ask you when Cosm should activate the trigger. This is where you define the threshold value or the rule that will activate the trigger. It can be one of these operators: >, >= , <. <= and ==. You can also select the 'any new value operator, which will take effect in case the datastream value changes, the 'frozen', or 'live' operators that will inform you incase your datastream becomes frozen or live respectively. Select the '<=' operator and set '300' for 'VALUE', so that are URL will be invoked when sensed light drops below 300, and click 'Next'.

The second parameter is the URL of your Arduino-based Web server. You set the full URL with the IP address as described previously, like: http://94.65.132.102/relay1. Finally, click the 'Create' button.

Now, you can test the effectiveness of your trigger without having to wait so that it gets dark in your room: Move you mouse over the trigger. Configuration options will appear on the right. Click the 'Debug' button (with the bug icon) on the page and watch after a few seconds the relay of your circuit to change state!

Use the Cosm APIs to visualize your data

One important feature of the Cosm service is allowing users to generate graphs that visualize the history of datastreams and use them in their own applications.

You can create such graphs in two ways:
1. Either get access to the history of the datastream and use your own methods to generate custom graphs and import them into your own applications. All methods that deal with the last 24 hours of historical data from single feeds are accessible from http://www.cosm.com/feeds/ and do not require authentication. For example, the URL: http://www.cosm.com/feeds/28602/datastreams/0/history.csv will return comma-separated values of the last 24-hour datastream updates from my room's temperature.
2. Use the Cosm API that generates history graphs and import them directly into your own applications.

The latter method is very convenient especially in cases you need to embed graphs into a website. The URL that generates PNG graphs also allows user to enter specific parameters regarding the size of the graph, to enter legends, set the title, etc.

For example the following link will generate the graph illustrated in Figure 8-11:

```
http://www.cosm.com/feeds/28602/datastreams/0/history.png?w=
600&h=300&c=33cc66&b=true&g=true&t=My%20Home%20Temperature&l
=My%20datastream&s=6&r=3
```

The following table presents the graph parameters that user can define as in the previous example:

Table 8-1. *Graph parameters that can be defined by user.*

Parameter	Description
w	Width
h	Height
c	Color
t	Title
l	Legend
s	Stroke size
b	Axis labels
g	Detailed grid
r	Resolution

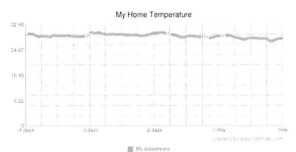

Figure 8-11. Example of the PNG graph generated by the Cosm API for visualizing a datastream.

Another useful feature of the visualization API is the generation of flash-enabled zoomable 30-day graphs. To do this you need initially to access a web application by Cosm and set up the appropriate parameters for your feed. The application for generating the graph is available at: http://apps.cosm.com/google_viz/

Initially you need to enter the Feed ID and the datastream ID you are interested in visualizing as in Figure 8-12. Then you need to set the size parameters of the graph (width and height) and the color as in Figure 8-13. By entering all the essential information, Cosm will generate a JavaScript code that can be embedded in your web site and illustrate a full 30-day zoomable graph of the datastream's history.

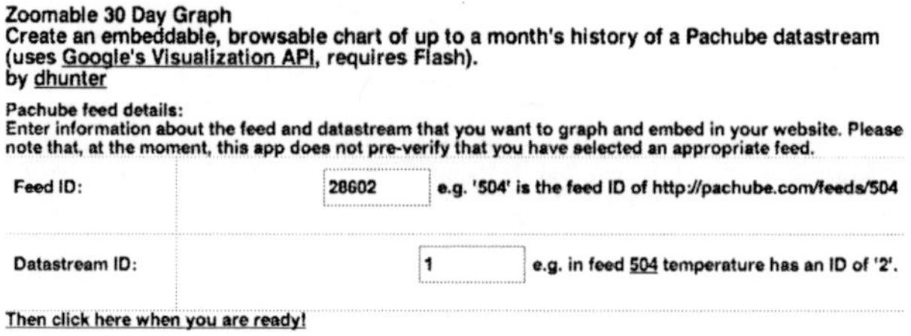

Figure 8-12. Configuring the 30-day zoomable graph

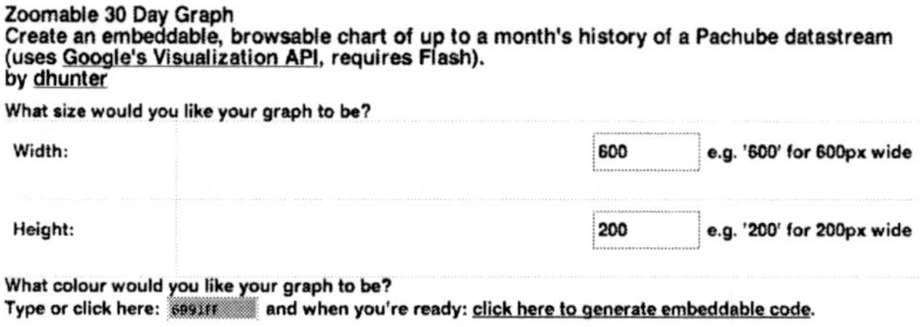

Figure 8-13. Configuring the 30-day zoomable graph

For example, entering the configuration parameters for feed 28602 and datastream 1 you get the following JavaScript code:

```
<script type="text/javascript"
src="http://www.google.com/jsapi"></script><script
language="JavaScript"
src="http://apps.pachube.com/google_viz/viz.js"></script><sc
ript
language="JavaScript">createViz(28602,1,600,200,"6991FF");</
script>
```

Adding the code into an html page will automatically generate the graph illustrated in Figure 8-14.

This is a preview of your embedded code:

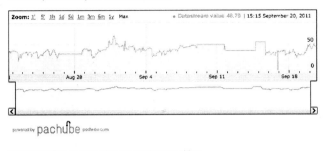

Copy this HTML to embed it in your webpage or blog:
```
<script type="text/javascript" src="http://www.google.com/jsapi"></script>
```

Figure 8-14. The generated 30-day zoomable graph with JavaScript code that can be directly embedded in your web site.

The Android Cosm Viewer

As a Cloud-based service, Cosm features viewing and retrieving feed and datastream data over a variety of platforms and applications. This is mainly achieved by the simple (web service-based) API they are using, the data format (JSON, XML or CSV) and the libraries that already exist for abstracting communication in several programming languages (like JPachube you have already used in this Chapter).

You have already been presented with how you can use datastream graphs in your own applications or websites through the Cosm visualization API. Cosm is also already supported by several mobile applications that allow users to view and edit feeds from the mobile device. One of them is the Android Cosm Viewer that enables viewing datastreams of any public feed, datastream history and feed location (see Figure 8-14 and Figure 8-15). The application can be downloaded from the Android market or here: http://code.google.com/p/pachube-android-viewer/. You can obtain the source code and modify it to allow users use a share key for viewing feeds and also edit feeds as well.

You can use the following code examples for creating, retrieving and updating feed data with the help of JPachube library:

Listing 8-10. Sample Android code for creating a new Cosm feed

```
//essential library imports:
import Pachube.Data;
import Pachube.Feed;
import Pachube.FeedFactory;
import Pachube.Pachube;
import Pachube.PachubeException;
....
```

```
try {
    //Create Cosm object authenicated using the MASTER
    //API KEY
    Pachube p = new Pachube("MASTER API KEY");

    //Create Feed object
    Feed mobilefeed = new Feed();
    mobilefeed.setTitle("My new feed");

    //Create datastream object and set id max and min valie,
    //tag and current value
    Data Pdatastream = new Data();
    Pdatastream.setId(0);
    Pdatastream.setMaxValue(100d);
    Pdatastream.setMinValue(0d);
    Pdatastream.setTag("Test");
    Pdatastream.setValue(30d);

    //Add datastream to feed
    mobilefeed .addData(a);

    //Retrieve the feed ID
    System.out.println(mobilefeed .getId());

 } catch (PachubeException e) {
    //If an exception occurs it will print the error message
    //from the failed
    System.out.println(e.errorMessage);
 }
```

To update a datastream's current value of a particular feed we can use the following:

```
try {
    Pachube p = new Pachube("MASTER API KEY");
    Feed feed = p.getFeed(28602);
    feed.updateDatastream(0, 28.2);
    } catch (PachubeException e) {
        System.out.println(e.errorMessage);
}
```

where 28602 is the feed ID, 0 is the datastream id and 28.2 the current datastream value (e.g., the temperature).

Figure 8-15. Screenshots from the Android Cosm viewer. Users can view any public feed.

Figure 8-16. Screenshots from the Android Cosm viewer. Users can also view the history of the datastreams and the location of the feed (if available).

Summary

This Chapter has introduced you to the Cosm Cloud-based Sensor Management. As you have seen it is a great service that allows users to manage online their sensor data, providing nice graphs and other visualizations of data history. It also allows you to set triggers on your sensor data and have Cosm service invoke a URL (e.g. a Web service of yours).

The Chapter provides the basics about the Cosm concepts and what are the steps you need to make in order to get started with it. The presented project demonstrates how to use Cosm in order to store and visualize temperature, humidity and light conditions of your room using either an Ethernet-enabled, or a WiFi-enabled Arduino or using your Android phone to forward the data to Cosm. It also instructs you how to use triggers in order to activate a relay switch based on your sensor readings.

Next Chapter talks about another famous sensor management platform, Nimbits.

9 INTRODUCING THE NIMBITS PUBLIC CLOUD SERVER

In the previous chapter you have been introduced with the popular Cosm Internet of Things platform that helps you manage online and visualize your sensor data. But what if you need to do more than that? In this chapter you will explore an alternative service, the Nimbits public Cloud Server. Nimbits is powered by the Google App Engine and apart from hosting and visualizing sensor data it allows users to install their own instance of Nimbits on a Cloud Infrastructure, to use all its features in standalone application through the Nimbits SDK and also to perform special operations on the sensor data like calculations and compression.

We will discuss all basic concepts of Nimbits service, demonstrate how you can use it to manage your sensor data online and also present an Arduino project that lets you log local weather data on the Cloud. Furthermore, you will utilize the intelligence of Nimbits in order to retrieve online information related to the sensor readings and we will also discuss the advanced features of Nimbits, like data compression, expiration and alert management. Finally, you will also check how you can deploy your own Nimbits sever utilizing the Google App Engine.

Introduction to basic Nimbits Concepts

Before you can use the Nimbits server for managing your own sensor data, it is necessary to become familiar with the basic concepts behind the platform that will also allow you to comprehend better its functionality and use Nimbits for your own projects.

Data Points
Nimbits, similar to Cosm, handles sensor data in 'Data points'. A Data Point is created within the Nimbits management environment and usually refers to a particular sensor and/or datastream (e.g., temperature). It is used to receive data through the Nimbits API and to visualize the data through the graphical environment. A data point can have several properties, like units of measure (e.g., Celsius or Fahrenheit), be public or private and a description about what the Data Point represents, how it is acquired, etc. Every Data Point in Nimbits environment is also associated with a QR code that you can use for easily retrieving the current value and history of the datastream using your mobile phone.

Data Point Categories

A simple but effective way of organizing your Data Points is through collecting them into categories. You can use the feature to organize your sensor readings in groups according the environment sensing (e.g., indoor, outdoor, etc.), the Arduino setup (e.g., Arduino1, Arduino2, etc.), the various projects (e.g., my temperature readings, my humidity readings, etc.) and so on.

Data Channels

The representation of a data point current value and history is performed through the Data Channels. A Data Channel is used to represent and visualize the values of the data points and additional information like the status of a data point (live or not), annotation values (like textual description of the data point), textual values, and timestamp values about that last data point update.

Authentication & Data Access

One major difference of the Nimbits service compared to other platforms like Cosm and ThingSpeak is that user authentication is performed using the Google accounts. Once you have already registered Google account credentials you can directly access and start using the Nimbits service without any additional registration/activation steps.

For authorized access to the Data Points Nimbits is following a similar authentication approach to the other platforms. By successfully authenticating for the first time into your Nimbits account, Nimbits generates a unique identification key, called the 'Secret Key' (a string of alphanumeric characters) that you can utilize in order to authenticate your devices when using the API for sending data to Nimbits.

Connections

'Connections' is a 'socializing' feature of Nimbits. You can use it to invite friends (having a Google account) who will be able to share their Data Points with yours. Based on permission levels you are able to see and interact with their Data Points and also allow them to do the same with yours.

The Web Service API

The most basic way of communicating either your Arduino board or your data management application with Nimbits is through the provided Web Service API. The API enables you to push and pull data from a Nimbits Server using a collection of Web Services that allow you to create data points, categories, and record data to your channels in many different ways.

It consists of simple POST and GET actions and responds with simple textual and/or JSON-formatted data.

Authenticating against the Web Service API (either you use in on Nimbits default server or through your own Google App instance) is performed using Google Accounts and the Nimbits secret key.

Connecting your Arduino with Nimbits

Let's see the basic features of receiving and logging sensor data in Nimbits Public Cloud server through an example: you will built a weather data logging system using Nimbits and your favorite Arduino board.

Log your Weather Data on the Nimbits Cloud

The purpose of the project is to demonstrate you how you can utilize the various features of Nimbits platform for logging and processing weather data on the Cloud. It consists of two basic steps: setting up and configuring your Nimbits environment for receiving the data and building the circuit that reads barometric data and forwards them to the Cloud using either a wired or a wireless network interface.

Set up Nimbits for receiving the Data

The first thing you have to in order to access Nimbits is to access the public Server Cloud environment at http://app.nimbits.com. To enter the environment you will need a valid Google account. After logging into your account you will be asked whether you with to allow Nimbits to use your account for authentication and then you will be redirected to the initial Nimbits start page that will look similar to the screenshot in Figure 9-1. This is also the main Nimbits interface that enables you to create and manage Data Points.

Figure 9-1. The main interface of Nimbits Public Cloud Server

The main Nimbits interface is divided in 4 basic sections. On the left, the vertical panel entitled 'Navigator' allows you to define and group your Data Points. The main window area on the server displays the data channels and the graphical charts that visualize the history of your sensor data. The vertical area on the left allows you to configure and manage remote Data Points through the 'Connections' you make. Finally, the bar on the top level of the screen contains various menu options for enabling instant messenger, Twitter and Facebook posts and controlling your main data access key.

To setup your Nimbits interface for logging your weather data you need to define the appropriate Data Points. Click on 'New Data Point' button under the 'Navigator' panel and create two new Data Points naming them after 'Temperature' and 'Pressure' respectively. The latter points will be used to store the data as generated by the barometric pressure as explained in the following section. As soon as each Data Point is created you will see it listed under the 'Navigator'. Next, click on the 'Category' icon (the one looking like a blue folder right next to the New Data Point icon) and give a name to your new category. Then, drag the two Data Points and drop them into the category so that you create the structure illustrated in Figure 9-2.

Figure 9-2. The structure of the Weather data Category

Note that the organization of the Data Points into categories does not serve any special functionality it just helps you organize your sensor data better!

DoubleClick on each Data Point to activate the respective Data channel. You will see in the central window the two channels that initially contain only the name of the point (no value or timestamp).

Before moving on to the circuit and the Arduino code you can check the communication with the Nimbits Cloud and the features of your datapoints by simply invoking the Nimbits API within your browser. To do so you need first to retrieve your secret key. Click on the 'Secret Key' icon and copy the 36-character key that is unique for your account and is required for storing datapoint values through the API.

Next, open the following URL on your browser by replacing the email with your Google account and by placing your own secret key at the end:

http://nimbits1.appspot.com/service/currentvalue?value=22&point=Pressure&email=arduino.sensors.cloud@gmail.com&format=double&secret=YOURSECRETKEY-GOES-HERE

The URL shall provide you back the sensor value you have just entered into your Data Point. Now go back to your Nimbits interface and double-click on the name of Data Point on the Navigator panel. You will see in the Data Channels window the value of the Data Point you just set. Click under the Value tab and enter a new value for your Data Point. You will notice that the graph visualizing the point's history gets automatically updated.

The next section will show you in practice how to use the Nimbits environment to log your Arduino data by reading weather data through the Barometer.

Reading Data from the Barometer

To acquire data from your environment and be able to make an estimation about the weather we need air pressure and temperature readings. The first is acquired usually by an instrument called barometer. If you can recall from Chapter 5 you have used such a sensor from Sparkfun in order to demonstrate the I2C digital communication with sensors and the Arduino. If you don't remember about it, just go back and have a look at it, the sensor board and the circuit setup diagram. You will be using the same sensor for this project.

You will need a BMP085 barometric pressure, an Arduino board and an Ethernet/WiFi shield (or an Ethernet/WiFi-enabled Arduino board).

The connection between the sensors and the Arduino board is pretty straight forward: you only need to connect analog pins 4 and 5 with SCL and SDA pins on the sensor respectively, power the sensor with 3.3V from the board and also connect the GND pins together. The connections should exactly as in Figure 9-3.

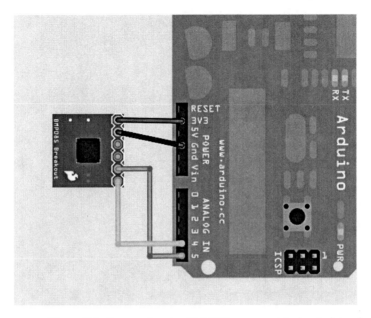

Figure 9-3. Connecting the BMP085 sensor with the Arduino board

As also discussed in Chapter 5, the communication with the Barometer sensor is not very straightforward and require some additional coding that implements and interprets the I2C protocol.

Listing 9-1. Functions for Callibrating the Module and Reading Temperature and Pressure Values. Code Adopted from Jim Lindblom's sketch (Sparkfun electronics)

```
void bmp085Calibration()
{
  ac1 = bmp085ReadInt(0xAA);
  ac2 = bmp085ReadInt(0xAC);
  ac3 = bmp085ReadInt(0xAE);
  ac4 = bmp085ReadInt(0xB0);
  ac5 = bmp085ReadInt(0xB2);
  ac6 = bmp085ReadInt(0xB4);
  b1 = bmp085ReadInt(0xB6);
  b2 = bmp085ReadInt(0xB8);
  mb = bmp085ReadInt(0xBA);
  mc = bmp085ReadInt(0xBC);
  md = bmp085ReadInt(0xBE);
}

// Read the temperature value
unsigned int bmp085ReadUT()
{
  unsigned int ut;

  // Write 0x2E into Register 0xF4
```

```
  // This requests a temperature reading
  Wire.beginTransmission(BMP085_ADDRESS);
  Wire.send(0xF4);
  Wire.send(0x2E);
  Wire.endTransmission();

  // Wait at least 4.5ms
  delay(5);

  // Read two bytes from registers 0xF6 and 0xF7
  ut = bmp085ReadInt(0xF6);
  return ut;
}

// Read the pressure value
unsigned long bmp085ReadUP()
{
  unsigned char msb, lsb, xlsb;
  unsigned long up = 0;

  // Write 0x34+(OSS<<6) into register 0xF4
  // Request a pressure reading w/ oversampling setting
  Wire.beginTransmission(BMP085_ADDRESS);
  Wire.send(0xF4);
  Wire.send(0x34 + (OSS<<6));
  Wire.endTransmission();

  // Wait for conversion, delay time dependent on OSS
  delay(2 + (3<<OSS));
  //Read register 0xF6 (MSB), 0xF7 (LSB), and 0xF8 (XLSB)
  Wire.beginTransmission(BMP085_ADDRESS);
  Wire.send(0xF6);
  Wire.endTransmission();
  Wire.requestFrom(BMP085_ADDRESS, 3);

  // Wait for data to become available
  while(Wire.available() < 3)
    ;
  msb = Wire.receive();
  lsb = Wire.receive();
  xlsb = Wire.receive();

  up = (((unsigned long) msb << 16) | ((unsigned long) lsb
  << 8) | (unsigned long) xlsb) >> (8-OSS);

  return up;
}
```

The first function (bmp085Calibration()) reads the calibration values form the device's EEPROM and stores them into program variables. It uses the bmp085ReadInt() function by passing the EEPROM data register addresses (from 0xAA up to 0xBE) as input parameters.

The last two functions read the voltage values that correspond to variations in measured temperature and pressure. As with the previous functions, they initialize the communication with the module defining its I2C address. They send the appropriate commands and wait the essential amount of time as defined by the datasheet. Then they send the byte values that correspond to the temperature and pressure reading using the bmp085ReadInt() function.

Listing 9-2. Using The Calibration Values to Calculate Actual Temperature and Pressure Readings

```
short bmp085GetTemperature(unsigned int ut)
{
  long x1, x2;

  x1 = (((long)ut - (long)ac6)*(long)ac5) >> 15;
  x2 = ((long)mc << 11)/(x1 + md);
  b5 = x1 + x2;

  return ((b5 + 8)>>4);
}

// Calculate pressure given up
// calibration values must be known
// b5 is also required so bmp085GetTemperature(...) must be
called first.
// Value returned will be pressure in units of Pa.
long bmp085GetPressure(unsigned long up)
{
  long x1, x2, x3, b3, b6, p;
  unsigned long b4, b7;

  b6 = b5 - 4000;
  // Calculate B3
  x1 = (b2 * (b6 * b6)>>12)>>11;
  x2 = (ac2 * b6)>>11;
  x3 = x1 + x2;
  b3 = (((((long)ac1)*4 + x3)<>2;

  // Calculate B4
  x1 = (ac3 * b6)>>13;
  x2 = (b1 * ((b6 * b6)>>12))>>16;
  x3 = ((x1 + x2) + 2)>>2;
  b4 = (ac4 * (unsigned long)(x3 + 32768))>>15;

  b7 = ((unsigned long)(up - b3) * (50000>>OSS));
  if (b7 < 0x80000000)
    p = (b7<<1)/b4;
  else
    p = (b7/b4)<<1;
```

```
x1 = (p>>8) * (p>>8);
x1 = (x1 * 3038)>>16;
x2 = (-7357 * p)>>16;
p += (x1 + x2 + 3791)>>4;

return p;
}
```

The complete code for reading the BMP065 module temperature and pressure values can be found here: http://www.sparkfun.com/tutorial/Barometric/BMP085_Example_Code.pde

Before attempting any data communication with the Internet, use the default BMP085 code example and verify that you receive proper readings through the Serial Monitor.

Sending Data using the Ethernet Shield

Let's see now how you can use your Ethernet shield to send temperature and pressure data to Nimbits server.

Listing 9-3. Using the Ethernet Library for storing Temperature and Pressure Data to Nimbits

```
#include <SPI.h>
#include <Ethernet.h>

//Define Ethernet variables here:
byte mac[] = {  0xDE, 0xAD, 0xBE, 0xEF, 0xFE, 0xED };
IPAddress server(72, 14, 204, 104); // Google Appspot
address

// Define the Ethernet client object
EthernetClient client;
...
//Initialize Ethernet module
// start the Ethernet connection:
if (Ethernet.begin(mac) == 0) {
    Serial.println("Failed to configure Ethernet using
    DHCP");
    // no point in carrying on, so do nothing forevermore:
    for(;;)
        ;
}
//give the Ethernet shield a second to initialize:
delay(1000);

void loop()
```

```
{
  temperature = bmp085GetTemperature(bmp085ReadUT());
  pressure = bmp085GetPressure(bmp085ReadUP());
  String stemp = String(temperature,DEC);
  String spress = String(pressure,DEC);
  //Attemp to connect and log data
  if (client.connect(server, 80)) {
    // Make a HTTP request for temperature:
    client.println("GET /service/currentvalue?value=" +
    stemp+"&point=Temperature&email=youracount@gmail.com&for
    mat=double&secret=YOURSECRETKEY HTTP/1.1");
    client.println("Host:nimbits1.appspot.com");
    client.println("Accept-Language:en-us,en;q=0.5");
    client.println("Accept-Encoding:gzip,deflate");
    client.println("Connection:close");
    client.println("Cache-Control:max-age=0");
    client.println();
    client.stop();
  }

  if (client.connect(server, 80)) {
    // Make a HTTP request for pressure:
    client.println("GET /service/currentvalue?value=" +
    spress
    +"&point=Pressure&email=youracount@gmail.com&format=doub
    le&secret=YOURSECRETKEY HTTP/1.1");
    client.println("Host:nimbits1.appspot.com");
    client.println("Accept-Language:en-us,en;q=0.5");
    client.println("Accept-Encoding:gzip,deflate");
    client.println("Connection:close");
    client.println("Cache-Control:max-age=0");
    client.println();
    client.stop();
  }

  //Make sure you get proper readings from the sensor
  //before sending them to Nimbits
  Serial.print("Temperature: ");
  Serial.print(temperature, DEC);
  Serial.println(" *0.1 deg C");
  Serial.print("Pressure: ");
  Serial.print(pressure, DEC);
  Serial.println(" Pa");
  Serial.println();
  //Send data every 20 seconds
  delay(20000);
}
```

Using Advanced Features

Compression

In case of Nimbits data compression means actually data filtering. It is a very interesting and unique feature that allows you to filter out any noise of the incoming sensor data. The idea behind 'compression' is that you are interested in and want to log considerable sensor changes (e.g., changes in temperature greater than 1°C) but the temperature sensor resolution is at 0.1°C. So instead of logging changes between 25.1 and 25.9 Nimbits will only log a new sensor value over 26.1°C.

When a new value is recorded to a Data Point, Nimbits first checks the compression setting. Compression tells Nimbits to only record values that are outside of the range of the compression setting, or if the incoming RecordedValue is significantly different than the previously recorded value. When you set a Data Point compression property to a value greater than zero, the system will determine if the newly RecordedValue is outside of the previously RecordedValue + or - the compression value. If a value is ignored do to a compression setting, all other services are skipped such as alerts.

To activate compression on your weather-logging project for the temperature Data Point, right click on the Temperature Data Point and click 'Properties'. Set the Compression value to 1.0 as in the following screenshot:

Figure 9-4. Setting the Compression value for the Temperature Data Point

You can turn compression off on a Data Point by setting PointCompression to -1.

Alerts

The Nimbits 'Alerts' feature can be compared to the 'Triggers' found in Cosm environment. Their functionality can be easily guessed by its name: whenever you need to be informed about an important change of a sensor's value, you set the appropriate alert. According to the sensor value, a data point can have four different states: normal, high (the last recorded value has been higher than or equal to the alert threshold), low (the last recorded value has been lower than the alert threshold) and idle (no changes in the sensor value). When a point goes into an Alert State, it will attempt to notify you in ways set in the point's properties. Currently, you can be alerted through email, instant messenger, Facebook and Twitter. In case you are browsing your Nimbits account through the Android mobile app the data points icon also change color according to the alert state.

Calculations

When a Data Point is written to with a new value, the event of recording a value can trigger a calculation.

The formulas (basic arithmetic) are executed and the resulting value is recorded into the target point – all compression, alerts and other functionality are then be applied to the new value as if it was a newly recorded value. A Data Point can be a trigger, parameter or target for a formula. If a point is a trigger and a new value is written to it the formula assigned to that data point will be executed.

After the formula executes successfully, the resulting double value is stored in the target point and all Record Value settings are processed (Such as Alerts, Compression etc.)

An Example

To setup a calculation you need to create at least two Data Point channels in Nimbits. A Trigger point contains the formula to be executed whenever that point received a new value. A target point receives the result of the calculation. You can also setup up to three parameter points. They will have their current values inserted into the calculation at variables x, y and z. Typically, the trigger point is also the x variable, causing the current value of the trigger (the newly recorded value) to be inserted into a formula and recorded to the target point. After creating your data points, log into Nimbits (app.nimbits.com) and double click the trigger point. You do not have to do anything to the target point. Select the point properties from the main menu and activate the calculation tab: Here, you can select the target point and points that should be the x, y and z values in the formula if you need them. Again, typically the target point is used as an x variable in the formula.

Enter a formula, ensuring the x, y and z are lower case. Click test to ensure everything is working.

Supported symbols are: , +, -, , /, ^, %, cos, sin, tan, acos, asin, atan, sqrt, sqr, log, min, max, ceil, floor, abs, neg, rndr. x + y | add the points stored in the x and y settings together and record the result in the target sqrt(x) | Take the square root of x and store in the target.

Data Expiration

Saving your sensor data online in Cloud infrastructures provides you with much more storage than you would have by using on your own resources so that you really do not have to worry about issues like extending physical hard disk or database capacity. However there can be cases that you might want your online data to be automatically deleted.

It can be good practice to configure a data point to automatically delete data that is older than a certain date. Depending on your application, you may be required to keep data up to a certain date.

Based on your point settings, Data can be permanently deleted when it reaches a certain age. Also, incoming data with a timestamp older than the expiration setting will be ignored by the system.

You can use a data point to record minute by minute data for a 24 hour period, and then store its 24 hour average into another point. By setting the minute to minute point's expiration to 24 hours, you can keep a record of 24 hour averages using very little storage.

Visualizing your Data on Nimbits Cloud Server

In Nimbits you can visualize the history and current values of your data points in two ways: a) using the Chart Web Service and b) using the JavaScript library for creating interactive charts.

The first option is more straightforward and easy to use when you simply want to integrate a graph into your own application. The service generates PNG images from your data. To do so it utilizes the Google Chart API. In the API you pass all the chart data in the URL through a POST HTTP request and the response is the generated image. For example, the following URL will generate the graph illustrated in Figure 9-5.

```
http://chart.apis.google.com/chart?cht=s&chd=t:12,87,75,4
1,23,96,68,71,34,9|98,60,27,34,56,79,58,74,18,76|84,23,69,81
,47,94,60,93,64,54&chxt=x,y&chs=200x125&chco=000000,ffff10
```

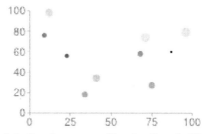

Figure 9-5. Graph generated by the Google Chart API

Nimbits has simplified the URL and all you need to add is the data point name and chart parameters. Nimbits will add the data for you. You can add the following parameters to the chart service request to tell Nimbits how to format your chart:

- point - a single point to add to the chart
- points - comma delimited list of points i.e. point1, point2, etc. for multiple trends
- count - a fixed number of values to add to the chart, current value and going back in time to the count
- sd - the start date of a date range to show, in Unix epoch milliseconds or a Nimbits timestamp
- ed - the end date of a date range to show, in Unix epoch milliseconds or a Nimbits timestamp

To see an example open your browser and add the following URL:

```
http://app.nimbits.com/service/chartapi?point=TempF&email
=bsautner@gmail.com&cht=lc&chs=300x200&chxt=x,y
```

you will see some temperature data from a temperature probe belonging to the Nimbits creator (Benjamin Sautner).

The second visualization method utilizes JavaScript in order to retrieve data points from Nimbits server in JSON format and visualize the latter using the Google Chart Tool. The provided JavaScript library takes as input the point name, the count for the number of values you'd like to pull down and the URL of the Nimbits server. The point is identified through a unique ID (UUID) that is unique to that point even across any number of Nimbits Servers that might be installed. To find about your point's UUID simply right click on the point (under the Data point Navigator), select Properties and then the Links tab.

For example, to make a simple HTML page that retrieves the value of a specific data point add the following code into it:

```
<html>
<head>
 <script type="text/javascript"
 src="http://app.nimbits.com/assets/js/nimbits
 3.3.0.js"></script>
 </head>
 <body>
 <a href="#" onclick="getPoint(callback,
 'http://app.nimbits.com', '96c55c29-d60f-42a1-82a9-
 98d8a76960bd', 10)">Click me!</a>

 <script type="text/javascript" language="javascript">
```

```
function callback(point) {
alert(point.name + " current value:" + point.value.d);
}
```

```
</script>
</body>
</html>
```

The JavaScript library is imported in the beginning of the html code and the function that retrieves the point is the getPoint() function. The UUID is 96c55c29-d60f-42a1-82a9-98d8a76960bd and the point is located on the default Nimbits server address (http://app.nimbits.com).

The idea behind the creation of dynamic charts is that you can combine the Google Chart API with the Nimbits JavaScript library for dynamically retrieving the data points instead of manually placing them inside the HTML code. The following example (code adopted from nimbits.com) demonstrates this functionality:

```
<html>
<head>
<script type="text/javascript"
src="https://www.google.com/jsapi"></script>
<script type="text/javascript"
src="http://app.nimbits.com/assets/js/nimbits-
3.3.0.js"></script>
<script type="text/javascript">
google.load("visualization", "1",
{packages:["corechart"]});
google.setOnLoadCallback(start);

function start() {
    getPoint(drawChart,'http://app.nimbits.com',
    '96c55c29-d60f-42a1-82a9-98d8a76960bd', 10)
}

function drawChart(_point) {
    var data = new google.visualization.DataTable();
    data.addColumn('string', 'Year');
    data.addColumn('number', _point.name);
    var values = _point.values;
    for (i = 0; i < values.length; i++) {
        var value = (eval('(' + values[i] + ')'));
        var date = new Date(value.timestamp);
        var ds = date.getFullYear() + "-" +
        date.getMonth() + "-" + date.getDate() + " " +
        date.getHours() + ":" + date.getMinutes() + ":"
        + date.getSeconds();
        data.addRow([ds, value.d]);
    }
    var options = {
```

```
            width: 400, height: 240,
            title: 'Nimbits Rocks!'
    };
    var chart = new
    google.visualization.LineChart(document.getElementById('
    chart_div'));
    chart.draw(data, options);
    }
    </script>
</head>
<body>
<div id="chart_div"></div>
</body>
</html>
```

The result of the code is the following graph:

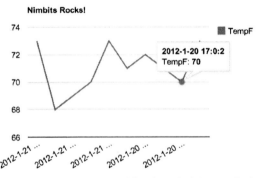

Figure 9-6. Dynamic graph generated by the Nimbits JavaScript library and the
Google Chart Tool

You can move your mouse over the graph plot and view information about
the specific data points.

For more information on Nimbits and data visualization your are
advised to check the Tutorials section at http://www.nimbits.com. You can
find examples on how to connect Excel spreadsheets with Nimbits,
generate Visio diagrams out of Nimbits data, barcodes and more. Nimbits
data export and visualization functionality is constantly updated and
extended!

Adding Intelligence to your Cloud Data

A great, useful and unique (as of writing this book) feature of Nimbits is the fact that you can add intelligence to your Cloud data. What does this really mean and how can it affect your data? Nimbits has recently integrated into WolframAlpha. For those not familiar with, WolframAlpha is an answer-engine developed by Wolfram Research (the company behind the *Mathematica* product). It features basically an intelligent online service that can answer textual queries (even in physical language) directly by computing the answer from structured data. WolframAlpha provides the intelligence behind the Apple's Siri and Android's Iris tools.

Nimbits has incorporated the WolframAlpha API allowing you to configure a data points to trigger a computation whenever it receives a new value. The result is then stored in a target Point. The result is that you can add intelligence to your software and hardware projects, using Nimbits' ability to relay, share and store your data streams and WolframAlpha's ability to understand and compute it.

Make a weather prognosis based on your elevation data

Let's see a quite simple but interesting example on utilizing the Nimbits Data Intelligence. We will extend the previous project with the BMP065 temperature and pressure sensor in order to make a simple weather forecast based on the sensed data. The idea behind this is that you can use the sensed air pressure to calculate an estimated altitude of your current location. From the estimated altitude you can calculate the expected air pressure. By comparing the estimated with the actual pressure from your sensor you can have an indication about what the weather will look like in the following hours.

To calculate the estimated altitude from the air pressure you can use the following formula:

$$\text{altitude} = 44330 * \left(1 - \left(\frac{p}{p_0} \right)^{\frac{1}{5.255}} \right)$$

where p stands for the calculated pressure and p_0 for the average pressure at sea level (101325 Pa).

The code for the later calculation is the following:

```
const float p0 = 101325;      // Pressure at sea level (Pa)
float altitude;
altitude = (float)44330 * (1 - pow(((float) pressure/p0),
0.190295));
```

Now you have to find out about your current location's altitude. While there are several ways to do so (using Google Maps, a GPS device, etc.), we will use Nimbits and WolframAlpha.

Go to your Nimbits instance and create a point called Location and another called Elevation. Enter your city name in the data column of the location field (e.g., Chicago) and click Save. It's also smart to add a value before trying to setup your WolframAlpha query, since you'll want some data to test with. Edit the properties of Location and give the following formula in the Intelligence Section:

Input: Elevation [Location.data]
Pod ID: Result
Result As: Text Data
Target Point: Elevation
Results in Plain Text: Checked.

Click Test Intelligence and you will get the altitude value back (e.g., 587 Feet for Chicago). Now go back and add a new city name. Click on save and wait for a few seconds. You shall see the Elevation point automatically updated to reflect the proper value.

This way you have yourself a data point that you can directly use it to retrieve your altitude within your Arduino code. To retrieve the point's value you can use the point's UUID and the Nimbits Web Service as explained in the previous section: check the point's properties and find the URL under the Links tab. It will look something like:

```
http://app.nimbits.com/service/currentvalue?uuid=dddcb0e2-
2fb4-4a6b-be43-daeba3fc2653
```

Next, to calculate the estimated pressure on this altitude you need to use this formula:

$$p = p_0 * (1 - \frac{altitude}{44330})^{5.255}$$

The code that connects to Nimbits, retrieves the Elevation data point, computes the estimated pressure and compares it with the actual one is the following:

```
IPAddress server(72, 14, 204, 104); // Google Appspot
address

// Define the Ethernet client object
EthernetClient client;
....
if (client.connect(server, 80)) {
    // Make a HTTP request for the Elevation point:
```

```
        client.println("GET /service/currentvalue?uuid=dddcb0e2-
        2fb4-4a6b-be43-daeba3fc2653 HTTP/1.1");
        client.println("Host:nimbits1.appspot.com");
        client.println();
        client.stop();
}
//Discard the HTTP header
readPastHeader(&client);

//start reading the response that contains the Elevation
//value
String elevation;
while (client.available()) {
        char c = client.read();
        elevation+=c;
    }

client.stop();


// convert the string value to float
char tmp[4];
elevation.toCharArray(tmp, 4);
float currentAltitude;
sscanf (tmp,"%f",&currentAltitude);


const float p0 = 101325;
// expected pressure (in Pa) at altitude
const float ePressure = p0 * pow((1-currentAltitude/44330),
5.255);
float weatherDiff;

// Add this into loop(), after you've calculated the
pressure
weatherDiff = pressure - ePressure;
if(weatherDiff > 250)
    Serial.println("Sunny!");
else if ((weatherDiff <= 250) || (weatherDiff >= -250))
    Serial.println("Partly Cloudy");
else if (weatherDiff > -250)
    Serial.println("Rain :-(");
Serial.println();
```

You will also need the following function in your code:

```
//The function that parses the HTTP response and discards
the HTTP header
bool readPastHeader(Client *pClient) {
  bool bIsBlank = true;
  while(true)
  {
```

```
if (pClient->available())
{
  char c = pClient->read();
  if(c=='\r' && bIsBlank)
  {
    // throw away the /n
    c = pClient->read();
    return true;
  }
  if(c=='\n')
    bIsBlank = true;
  else if(c!='\r')
    bIsBlank = false;
  }
}
}
```

The whole idea and process is adopted from the Sparkfun tutorial: http://www.sparkfun.com/tutorials/253. Special thanks to Benjamin from Nimbits for helping me to build this scenario using Nimbits.

Another great tutorial is the Intelligent M2M tutorial found on the Nimbits site. It demonstrates how to use the Nimbits Intelligence among with inter-device communication using the Instant Messaging protocol – XMPP.

Your Nimbits Data on your Android Phone

Nimbits offers also an Android app that you can use to control your data points, view alerts and data history graphs real time. You can download the app from the Android Market for free and use it both on a phone or an Android-enabled Tablet (see Figure 9.7).

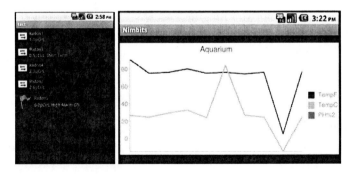

Figure 9-7. The Android Nimbits app. Works both on phones and Tablets providing information about data point and visualizing history graphs

Deploying your own Nimbits Server

As already discussed one of the great Nimbits features is the fact that you can download the server application and deploy it on your own Cloud environment.

Deploying your own instance of Nimbits Server involves downloading the latest version, configuring it and then deploying it to Google App Engine using a program that comes with the App Engine SDK called appcfg. Let's see the required steps into details:

First you need to log into App Engine with your Google account: https://appengine.google.com. Then click the "Create an Application" to create the application identifier that you will upload Nimbits to. Call it whatever you like. This will be your App ID and part of the URL you'll use to access your server. For example: yourAppID.appspot.com Download the latest version of the JAVA App Engine SDK for your platform and unzip into a folder on your computer. Download the Nimbits Server and extract the files to a folder. Edit the file Nimbits/server/war/WEB-INF/appengine-web.xml and replace the line below with your application ID that you created above.

You will now use the app engine SDK to deploy Nimbits to app engine. You can learn more about this process here:http://code.google.com/appengine/docs/java/gettingstarted/uploading.html

Simply run at a command line: appengine-sdk-directory/bin/appcfg.sh (or appcfg.cmd if you're using windows) from your App Engine SDK download directory and provide the location of the war directory from your Nimbits Server Download.

Google will prompt you for your email and password during the upload.

This command reads the appengine-web.xml you edited above and deploys it to app engine.

Return to https://appengine.google.com/ and select your Nimbits application ID. Under the Administration section select Versions. Make the app engine version number you just deployed is the current / default version.

Initialize your Nimbits Server (First Time Deploying)

Before you can use the server, the System Maintenance service must be ran once. Browse to this URL, replacing app.nimbits.com with your applicationID.appspot.com URL:

http://yourappid.appspot.com/cron/SystemMaint

Add your settings to the Data Store. Then return to https://appengine.google.com/ and select your Nimbits application ID. Click on "Data Store Viewer" under the Data category on the left. On the

Query Tab, select Server Settings. You should see the default settings that were generated when you ran the system maintenance service. You can click on these settings and edit their values to fit your needs. The admin email must be your account's email in order for email alerts being sent from the system to work. You can now use your own copy of the Nimbits Server!

Summary

Managing your sensor data online requires an appropriate web-based application that will provide the essential interfaces for communicating with your microcontroller board and will visualize your sensor data history. The Nimbits Public Cloud Server offers much more than that since it additionally provides advanced features like filtering sensor values through data compression, creating alerts and notifications through social networks (like Twitter and Facebook) and sharing sensor data through your friends. It also provides users with a flexible and easy to use API for reading and storing sensor values through any kind of custom application. Finally, it incorporates a data intelligence service that can be utilized for providing you with information related to your sensors and assist you in creating alerts and making decisions.

Nimbits is open source, meaning that you can download and modify the Java code to meet your needs or simply deploy it on a custom Cloud infrastructure or your own Google App Engine account.

Next Chapter will demonstrate you how to utilize such a Cloud infrastructure in order to reprogram your Arduino remotely.

10 REPORGRAM YOUR ARDUINO REMOTELY FROM THE CLOUD

No matter if you have programmed an Arduino board before reading this book, or your first attempt has been by reading Chapter 4, you must have noticed one think: the board has to be attached on your computer's USB port for uploading your sketch. The slightest change in the code (either because you notice a logic flow error or you need to modify/add functionality) requires that you connect the Arduino with your computer (and in most cases you disassemble it from the rest of the circuit).

While this process is not usually a big issue when you develop the code for your project and test it, it is not that convenient when you already have deployed your project. Especially in the case of IoT projects where you might have established remotely several boards connected to various sensors reporting data on the Cloud. The solution to this is to program your Arduino remotely.

This Chapter will demonstrate you how to reprogram your Arduino remotely using a network interface and a web application on the Cloud. It starts by briefly discussing how the Arduino is programmed. Then alternative methods for programming the Arduino wirelessly over ZigBee or Bluetooth is presented.

How is the Arduino Programmed

When you click the upload button on the Arduino IDE there are several background processes that take place before your board starts to execute your sketch code. The code itself needs to be verified for errors (syntactical errors), and then compiled and transformed into binary format so that it can be transferred to and executed by the ATmega328 microcontroller processor on your board.

Before the actual upload of the binary file, the process of code building takes place. During the latter the Arduino IDE performs four major tasks using appropriate tools:

- Combines Files: Your sketch code might be consisted of several different .pde files that need to be concatenated together.
- Transformations: Adds function prototypes and #include statements.
- Compiles Code: Each .c and .cpp is compiled into .o files.
- Linking: The .o files are linked together with library files.

The result of these tasks being completed is a .hex (binary) file that is ready to be uploaded to the Arduino hardware using a special program called 'avrdude', a very popular command line chip for programming AVR chips.

The upload process is also controlled by variables set by the IDE that have to do with:

- the protocol that avrdude should use to talk to the board (typically "stk500").

- the speed (baud rate) avrdude should use when uploading sketches (typically "19200").

- the maximum size for a sketch on the board (dependent on the chip and the size of the bootloader).

The file combination, transformation, compilation and linking can be performed manually, using the appropriate compiler and compilation parameters. If you wish to view all these intermediate steps that take place when you compile your sketch within the Arduino IDE, simply edit your IDE properties and enable 'Show verbose output during' for both 'compilation' and 'upload' options like in the following figure:

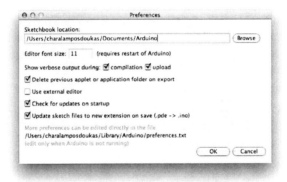

Figure 10-1. Setting the Arduino IDE parameters for verbose sketch compilation and uploading

Now open a simple sketch in your Arduino IDE and click on 'Verify' or 'Upload' (assuming you have your board connected) button. You will notice that the IDE prints out all the commands that are executed in the background that compile, link your code, prepare the .hex file and upload it:

Figure 10-2. Setting the Arduino IDE parameters for verbose sketch
compilation and uploading

The programs used for the latter processes are: avr-gcc, avr-j++, avr-ar, avr-objcopy and avrdude.

The avrdude follows a specific process for uploading the code (part of the "stk500" protocol mentioned previously). The description of the protocol can be found here: http://www.atmel.com/Images/doc2591.pdf. The following steps are a very brief overview of the microcontroller chip programming process:

- Reset the chip
- Send syncing commands
- Send programming parameters
- Enter programming mode
- Write the program code in block
- Leave the programming mode

All steps are implemented by the avrdude program.

The communication between your computer and the board is made through USB using the Serial protocol (as every other communication for sending and receiving data to and from the board when your sketch is executing). So in order to program your board all you need is the final binary code (.hex file), a serial link with the Rx/Tx pins of the board and the avrdude application. Obviously the only way to execute the avrdude is through a computer, therefore one option to wirelessly program the board is to substitute the USB link though a wireless interface (e.g., ZigBee or Bluetooth that provide a serial interface to your computer). This is discussed in the following Section.

The only way to completely reprogram you Arduino board without the need for a computer is to reproduce the programming performed by the avrdude application using another microcontroller, e.g., another Arduino. This will be presented also in this chapter among with a web application on a Cloud infrastructure that allows you to upload your sketch's code.

Programming Remotely Using ZigBee

A very useful tutorial on how to program your Arduino remotely using a pair of ZigBee modules is provided by Limor Fried at: http://www.ladyada.net/make/xbee/arduino.html

The basic idea is that you replace the USB serial link with the board using a ZigBee wireless connection. This requires two ZigBee nodes, one attached to your computer and one attached on the Arduino board to be reprogrammed. You need to provide power and ground connections to the ZigBee module from your board and connect the Rx/Tx pins of the module to the Tx/Rx pins of your Arduino respectively. In addition you need to connect the reset pin of the ZigBee module with the reset pin of your board so that you can wirelessly reset the board and begin the programming process.

The circuit setup looks like the one presented in Figure 10-3. You need to have one ZigBee module attached to the USB port of your computer and the other one to your board.

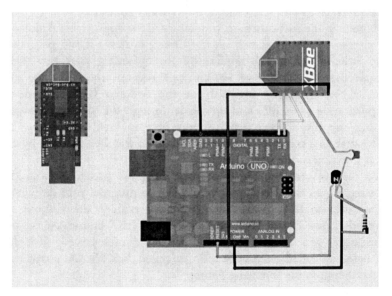

Figure 10-3. Reprogramming an Arduino board using ZigBee modules

While you can directly connect the reset pins, you are advised to follow the instructions by Limor Fried and interpret a small circuit in between (that uses a NPN transistor).

Another great idea of programming an Arduino board over a WiFi network with a help of a WiFi router is described here: http://www.elcojacobs.com/programming-my-arduino-over-wifi/

Programming Remotely Using an Internet Connection

This project presents you with everything you need for building a circuit that allows you to remotely program a target Arduino board using a Web-based application on the Cloud.

What you will need

In order to build the project you obviously need an Arduino with a wireless/wired interface. The wireless interface shall preferably provide direct access to the Internet, so it can be either WiFi or GRPS. The Internet-enabled Arduino shall be the 'master' Arduino that will be responsible for receiving the sketch code from the cloud and programming the 'slave Arduino. The latter can be a complete board or a standalone microcontroller chip as presented in Chapter 4. It will also be responsible for collecting any sensor information available and sending it to the master Arduino which in turn will forward it to a web application.

Regarding the software part, you will need the appropriate Arduino code that checks for a new sketch on the Cloud, retrieves it and programs the slave Arduino and a web application for uploading your new sketches. Because the compilation of sketch code requires tools like the avr-gcc compiler, for simplicity you will use the Arduino IDE to retrieve the compiled code (.hex file) and upload it through the web application. For this purpose the chapter demonstrates how to build and deploy a Java Servlet that also converts the binary format of the .hex file into byte code that can be sent to the slave Arduino.

To retrieve the .hex file from your compiled sketch you can do the following. Open the sketch you would like to program your Arduino with in the Arduino IDE. Make sure you have enabled the verbose sketch compilation (see Figure 10-2) and click the 'Verify' button. At the end of the messages you will notice the full path of the temporarily (it is deleted after being uploaded on your board) generated .hex file. On a mac it looks like the output in the following figure:

```
Done compiling.
/var/folders/td/_8bbgphd6d3b_cf10yvgts0w0000gn/T/build4186910207013183326.tmp/Blink.cpp.elf
/var/folders/td/_8bbgphd6d3b_cf10yvgts0w0000gn/T/build4186910207013183326.tmp/Blink.cpp.hex
```

Figure 10-4. The location of the generated .hex sketch file

The Circuit Setup

All you need for this project is to connect your Arduino boards according to the following figure.

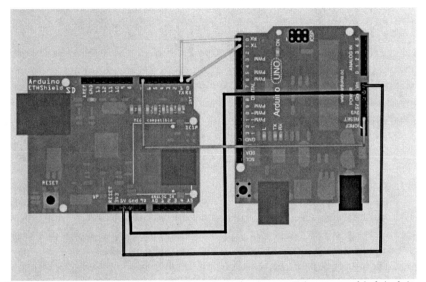

Figure 10-5. Reprogramming an Arduino board using an Ethernet-enabled Arduino

The master board is on the left and has also an Ethernet module (it can be WiFi as well) through an extension shield. I assume that the master board powers the slave board, so the Ground pins are connected the 5V output pin from the master goes to Vin pin of the slave. If you plan to separately power the slave board, make sure the Ground pins of the two board still remain connected. You need to connect the Rx/Tx pins of the master board with the Tx/Rx of the slave board respectively so that the communication between the two boards over the Serial protocol can take place. Finally, you need to connect the digital pin 7 of the master board (the pin selection is code-based and can be changed if you wish) with the reset pin of the slave board.

Deploying your Servlet on the Cloud

As with all the web applications presented in the book, this one will be also hosted on a Cloud platform. The main requirements of the web application is that it can create new files on the hosted environment (for the new sketch files that you will upload) and that it can connect to a MySQL database (so that it can keep track of the new sketches submitted).

The Google App Engine has some limitations regarding the creation of file and no database support, so I have chosen a different Cloud application hosting environment for this project, the Jelastic Cloud Service.

The Jelastic Cloud Service

The Jelastic Cloud Service (http://www.jelastic.com) is a great PaaS (Platform as a Service) Cloud. It allows users to deploy Java-based applications providing all the essential components (application server instances, databases, load balancers, etc.) and all the appropriate scalability. Jelastic provides full access to the application server runtime environment, which enables the deployment of additional Java extensions like encryption and authentication libraries and most importantly: it allows applications to create and have access to local files. This last feature is quite essential for deploying our remote-programming application. Jelastic makes the best candidate.

Set up the Environment

First you need to signup (at http://jelastic.com)to the Jelastic environment (free service) in order to be able to use it. After completing the registration process you will be provided with the essential credentials to enter the service. When signing up you are asked to select a hosted service provider (either ServInt or HostEurope). I suggest you go with the first one so that the service addresses (URLs, database connections, etc.) match the ones described in the book and included in the code files.

The very first step is to create and configure the Jelastic environment for your application. For the purpose of the project you will need an application server (Tomcat 6) and a Database server (MySQL 5.0).

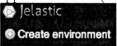

In your main Jelastic page click on the 'Create Environment' button and then configure your new environment by selecting the Tomcat 6.0 application server, MySQL 5.0 Database and add two servers as in the following image.

Give your new environment whatever name you like, I have chosen 'myarduino' for this purpose. Once done, click on Create and wait until Jelastic runs its magic on the background and sets up your environment.

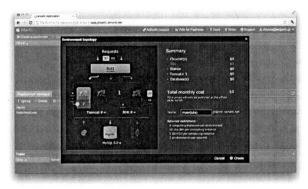

Figure 10-6. Setting up your Jelastic environment

In a couple of minutes the environment will be up and running and you will also receive an email notification about that, followed by an email that contains your login credentials for the MySQL environment and the URL to the phpmyadmin page for managing databases.

Next comes the setup of the database. Using the link on the email (will look like: http://mysql-myarduino.jelastic.servint.net) and the credentials log into phpmyadmin.

Go under 'Databases' tab and create a new database for managing the sketches. Create a new database and name it after 'arduino'. Do not worry about the collation, since we are storing simple values.

Then move on and add a table named 'sketch' with the attributes 'new', 'filename' and 'date' as follows.

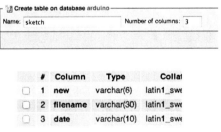

If you wish you can also create a new user with read/write permissions on the database so that you can use its credentials in your code. I have created a user 'arduino' with password 'arduino!' for this purpose.

Before moving one, make sure you make one initial entry in the sketch table. Just use some dummy data to enter. The record will be updated with true values by the file upload Java servlet.

Now you are done with the initial environment setup. Next thing you need to do is develop the web application and deploy it!

The Java Servlets

Let's start first with the code of the Java Servlets. The web application will be responsible for receiving the compiled sketch code (.hex) file from you and converting it to a suitable format so that it can be programmed on the target board. It will also notify the master Arduino board that there is a new sketch available so that the latter can download it and program the target board.

So for this project purposes you need to develop three Java servlets. The first one will handle file upload requests through a web form, the second one will be used to notify the master board for a new sketch and the third one will convert and provide the master board with the binary sketch code. (Depending on your Java skills you can easily migrate all three into one servlet, for simplicity and clarification I have developed each feature into a separate servlet).

The whole project code including all essential libraries and html file can be found in the book's online code repository. Code explanation follows for each file.

Upload.java

This servlet (as the name suggests) handled the file uploads from a web form (upload.html under the WebContent folder). The code starts with importing the appropriate library files for building the servlet, making the connection and queries to the MySQL database and for uploading files (org.apache.commons.fileupload.*).

The init() function is automatically invoked when the servlet is initialized and creates the appropriate directory for handling the uploaded sketch files into the user's home directory.

```
public void init(ServletConfig config) throws
ServletException {
    super.init(config);
    new File(path+"/mydir").mkdir();
    tmpDir = new File(path+"/mydir");
    if(!tmpDir.isDirectory()) {
        throw new ServletException(TMP_DIR_PATH + " is not a
        directory");
    }
    String realPath =
    getServletContext().getRealPath(path+"/mydir/");
    destinationDir = tmpDir;
    if(!destinationDir.isDirectory()) {
    throw new ServletException(realPath+" is not a
    directory");
```

```
    }
}
```

The doPost() method handles the POST requests issued by the client (the web browser) when the file upload is requested. For this purpose the ServletFileUpload extension by the apache commons library has been used. When a file is submitted, it is written on user's home directory in the Jelastic environment under the 'mydir' directory, and then updates the record in the 'sketch' table of the MySQL database setting the 'new' flag to 'true' and updates the sketch's filename and submission date.

The current implementation does not keep any record of the submitted sketches, it is only used to notify the Arduino master board (through the 'new' flag) that there is a new sketch code available. With minor modifications on the code and the database you can implement the essential functionality for handling more than one sketch files.

Check.java

This servlet provides a simple interface for the Arduino that allows it to check for new sketches on the Cloud server. The Arduino invokes the URL of the service and receives a sketch status as a response. When the servlet receives a GET request (a simple request to the URL of the service) it performs a query in the MySQL database and retrieves and prints status of the 'new' flag to the client output. This way the Arduino will get either 'true' or 'false' as a response from the server depending on whether there is a new sketch available or not.

Sketch.java

This servlet provides also a GET interface for the Arduino that enables it to request the new sketch file properly encoded. It takes the sketch's file name as an input from the 'file' parameter as follows:

```
http://myarduino.jelastic.servint.net/Sketch?file=Blink.cpp.
hex
```

then it looks for the file in the sketch file directory (mydir) within the user's home directory and reads the latter line by line. Then it needs to convert the binary file into a format that contains the byte representations so that the latter can be transferred to the Arduino. To do so the code starts a new conversion thread for each line. When all threads are completed the output is served into the client as plain text.

Upload & Test the Service

In order to deploy the web application you need first to build the code and export it into a deployable war package. You can use the Eclipse's export tool. Once you have the war file created go under the Jelastic Deployment manager and click the Upload button.

Select the file from the dialog, add a description for the war file and click the Upload button. Once done, you will see the application package under the deployment manager. Before testing the service you need to add the appropriate library files to the application server's (i.e., Tomcat) path. To do so click the configure icon next to Tomcat's icon.

The configuration window opens. Under the lib directory click the upload button to upload the essential library files: commons-fileupload-1.2.2.jar, commons-io-2.1.jar and mysql.jar. All three jar files can be found in the lib folder of the project's source code.

Once completed make sure you also restart the Tomcat node. Then go back to the deployment manager window, and deploy the application using the deploy icon:

You can leave the default context path (ROOT) or give a new one. When the process of package deployment is completed you will see the application listed under the Tomcat instance.

Next step involves the testing of the application. All you need to do is to invoke the web application by visiting the link in your browser which shall look something like:

```
http://myarduino.jelastic.servint.net/
```

You will see the form for uploading the hex file of a sketch. Compile a sketch in your Arduino IDE and retrieve the hex file as indicated previously and upload it to the java application. Then retrieve the binary form of the hex file by entering the following URL:

```
http://myarduino.jelastic.servint.net/Sketch?file=Blink.cpp.
hex
```

You shall see on your browser the output of the binary file.

The Arduino Code

The Arduino Code is responsible for checking for new sketches on the Cloud server. In case there is a positive response the code then invokes the URL that contains the byte representation of the sketch file. It reads the byte code in segments of 128 bytes and sends the latter to the target chip.

You can find the Arduino code in the book's online code repository. The provided code contains all the essential code and functions for communicating with the Cloud service and programing a target Arduino (AVR) microcontroller. A brief explanation of the sketch follows. The code starts with importing the essential libraries for Ethernet communication. It also includes the pgmspace library by AVR that includes the essential methods for programming the AVR microcontroller.

```
#include <SPI.h>
#include <Ethernet.h>

#include <avr/pgmspace.h>
```

Then various variables are defined that will be used to handle data.

```
#define BYTE 0

int began = 0, groove = 50, i, j, start, end, address,
laddress, haddress, error = 0, a, b, c, d, e, f, buff[128],
buffLength, k, readBuff[16], preBuff, readBuffLength;
String filename = "";
```

Among the variables the reset pin is also defined. This should be connected to the Reset pin on the target board:

```
int resetPin = 7;
```

Next come the variable definitions (like the MAC and server address) for the Ethernet communication:

```
byte mac[] = {  0x00, 0xAA, 0xBB, 0xCC, 0xDE, 0x02 };
char serverName[] = "arduino.jelastic.servint.net";
```

Then an EthernetClient object is created for handling the Internet connection to the server:

```
EthernetClient client;
```

The setup function performs all the essential initialization for the Ethernet setup and then invokes the ProgramArduino() function.

```
void setup() {
  // start the serial library:
  Serial.begin(57600);

  //start the Ethernet connection:
  if (Ethernet.begin(mac) == 0) {
    Serial.println("Failed to configure Ethernet using
    DHCP");
  }
  // give the Ethernet shield a second to initialize:
  delay(10000);
  Serial.println("connecting...");
  ProgramArduino();

}
```

Before programming the microcontroller chip you need to initialize the programming process by resetting the chip and sending an appropriate sequence of bytes that instantiates the process. These steps are performed by the initializeProgramming() function:

```
int initializeProgramming() {
```

The function initially sets the Serial rate at 57600 (default rate for programming the AVR chips) and also sets the reset digital pin in output mode so that it can generate the reset pulse:

```
  Serial.begin(57600);
  pinMode(resetPin, OUTPUT);
```

It then resets the target chip by sending first a LOW and then a HIGH pulse:

```
  digitalWrite(resetPin, LOW);
  delay(100);
  digitalWrite(resetPin, HIGH);
  delay(100);
```

and synchronizes with the AVR chip by sending an appropriate byte sequence as defined by the stk500 protocol:

```
for (i = 0; i < 25; i++) {
  Serial.write(0x30); // STK_GET_SYNC
  Serial.write(0x20); // STK_CRC_EOP
  delay(groove);
}
```

In order to make sure that the synchronization with the chip has been achieved, you need to check the response from the chip. This is performed by the readBytes() function that reads the response bytes into a global array named readBuff[].

```
readBytes();
```

After synchronization we expect to receive 0x14 (STK_INSYNC) and 0x10 (STK_OK) so the code checks for these responses:

```
if (readBuffLength < 2 || readBuff[0] != 0x14 ||
readBuff[1] != 0x10) {
    return 0;
}
```

It flushes the Serial ports (just in case):

```
Serial.flush();
```

Then the programming parameters are set according to the protocol:

```
Serial.print(0x42, BYTE);
Serial.print(0x86, BYTE);
Serial.print(0x00, BYTE);
Serial.print(0x00, BYTE);
Serial.print(0x01, BYTE);
Serial.print(0x01, BYTE);
Serial.print(0x01, BYTE);
Serial.print(0x01, BYTE);
Serial.print(0x03, BYTE);
Serial.print(0xff, BYTE);
Serial.print(0xff, BYTE);
Serial.print(0xff, BYTE);
Serial.print(0xff, BYTE);
Serial.print(0x00, BYTE);
Serial.print(0x80, BYTE);
Serial.print(0x04, BYTE);
Serial.print(0x00, BYTE);
```

```
Serial.print(0x00, BYTE);
Serial.print(0x00, BYTE);
Serial.print(0x80, BYTE);
Serial.print(0x00, BYTE);
Serial.print(0x20, BYTE);
delay(groove);
```

Again the code checks for the proper response from the AVR chip:

```
readBytes();
if (readBuffLength != 2 || readBuff[0] != 0x14 ||
readBuff[1] != 0x10) {
   return 0;
}
```

And continues with setting the extended programming parameters

```
Serial.print(0x45, BYTE);
Serial.print(0x05, BYTE);
Serial.print(0x04, BYTE);
Serial.print(0xd7, BYTE);
Serial.print(0xc2, BYTE);
Serial.print(0x00, BYTE);
Serial.print(0x20, BYTE);
delay(groove);
```

Another check for proper response:

```
readBytes();
if (readBuffLength != 2 || readBuff[0] != 0x14 ||
readBuff[1] != 0x10) {
   return 0;
}
```

The last step in the programming procedure initialization involves entering the programming mode by sending the appropriate BYTE flags to the chip:

```
Serial.print(0x50, BYTE);
Serial.print(0x20, BYTE);
delay(groove);
```

and checking for the proper response from the chip:

```
readBytes();
if (readBuffLength != 2 || readBuff[0] != 0x14 ||
readBuff[1] != 0x10) {
   return 0;
```

The Arduino chip is now ready to accept the new code. The new code is retrieved from the web application on the Cloud and the whole process is performed in the ProgramArduino() function as follows:

```
int ProgramArduino() {
```

The code in the ProgramArduino() function starts with checking for a valid connection with the server and makes a GET HTTP request for the retrieving the BlinkWithoutDelay.cpp.hex binary code.

```
if (client.connect(serverName, 80)) {
   // Make a HTTP request:
   client.println("GET
   /Sketch?file=BlinkWithoutDelay.cpp.hex HTTP/1.0");
   client.println("Host:arduino.jelastic.servint.net");
   client.println();
```

When the GET request is completed, the code needs to parse the response from the web application. Since the response contains additional information (like the HTTP header), a special function is used that removes such info. When done, it invokes the programming initialization procedure that has been previously implemented.

```
readPastHeader(&client);
initializeProgramming();
```

As soon as the header is parsed, the rest of the web response contains the string representation of the byte code for the new sketch file. Each code byte is represented as 0xff separated with a comma from the following byte. The code reads each character available and stores it into a temporary buffer[] array.

```
while (client.available()) {
   char c = client.read();

   if(charcounter <4) {
     buffer[charcounter]=c;
   }
   charcounter++;
```

When the buffer is full with 4 characters, the fifth character is skipped (because it is the comma that separates the byte values) and the whole byte represented by the 4 characters is stored into the tmpbyte char variable. The latter is then stored into the n[] array that holds all the parsed code bytes as received by the Cloud application.

```
if(charcounter==4) {
  //skip 5th char (,)
  client.read();
  total++;
  charcounter=0;

  if(bytecounter < 128) {
    char *tmpbyte = buffer;
    sscanf (tmpbyte,"%x",&n[bytecounter]);
    bytecounter++;
  }
```

If however the total parsed byte counter has reached 128, the code needs to send the byte data to the target chip:

```
else {
    bytecounter = 0;
    //Got 128 bytes let's program them
```

It sets the address of the AVR flash memory to write to:

```
haddress = address / 256;
laddress = address % 256;
address += 64; // For the next iteration
Serial.print(0x55, BYTE);
Serial.print(laddress, BYTE);
Serial.print(haddress, BYTE);
Serial.print(0x20, BYTE);
delay(groove);

readBytes();
if (readBuffLength != 2 || readBuff[0] != 0x14 ||
readBuff[1] != 0x10) {
  return 0;
}
```

and writes the block of 128 parsed bytes that are in the n[] array:

```
Serial.print(0x64, BYTE);
Serial.print(0x00, BYTE);
Serial.print(buffLength, BYTE);
Serial.print(0x46, BYTE);

for (j = 0; j < 128; j++) {
  Serial.print(n[j], BYTE);
}

Serial.print(0x20, BYTE);
```

```
        delay(groove);

        readBytes();
        if (readBuffLength != 2 || readBuff[0] != 0x14 ||
        readBuff[1] != 0x10) {
           return 0;
          }
```

At the end of the process, depending on the sketch size, there can be some bytes left that are needed to program them as well, following the similar process:

```
int bufflength = bytecounter;

// Set the address of the AVR flash memory to write to
haddress = address / 256;
laddress = address % 256;
address += 64; // For the next iteration
Serial.print(0x55, BYTE);
Serial.print(laddress, BYTE);
Serial.print(haddress, BYTE);
Serial.print(0x20, BYTE);
delay(groove);

readBytes();
if (readBuffLength != 2 || readBuff[0] != 0x14 ||
readBuff[1] != 0x10) {
   return 0;
}

// Write the block
Serial.print(0x64, BYTE);
Serial.print(0x00, BYTE);
Serial.print(buffLength, BYTE);
Serial.print(0x46, BYTE);

for (j = 0; j < bufflength; j++) {

  Serial.print(n[j], BYTE);

  }

Serial.print(0x20, BYTE);
delay(groove);

readBytes();
if (readBuffLength != 2 || readBuff[0] != 0x14 ||
readBuff[1] != 0x10) {
   return 0;
}
```

Finally the code leaves programming mode:

```
Serial.print(0x51, BYTE);
 Serial.print(0x20, BYTE);
 delay(groove);
 readBytes();
 if (readBuffLength != 2 || readBuff[0] != 0x14 ||
 readBuff[1] != 0x10) {
   return 0;
 }
```

After checking for proper response from the AVR chip, the code closes also the connection with the remote server:

```
client.stop();
```

The check() function that follows invokes the appropriate web service from the Cloud application and checks whether there is a new sketch available for download. It reads the response from the server and checks if the latter starts with "OK" which means that there is a new sketch available for download and also retrieves the sketch file name.

```
boolean check() {
String result = "";

if (client.connect(serverName, 80)) {
    //Serial.println("connected for check");
    // Make a HTTP request:
    client.println("GET /Check HTTP/1.0");
    client.println("Host:arduino.jelastic.servint.net");
    client.println("Accept-Language:en-us,en;q=0.5");
    client.println("Accept-Encoding:gzip,deflate");
    client.println("Connection:close");
    client.println("Cache-Control:max-age=0");
    client.println();
    readPastHeader(&client);
 }
 else {
 //...
 }

 while (client.available()) {
     char c = client.read();
     result +=c;
 }

 client.stop();
 if(result.startsWith("true")) {
     filename = result.substring(6, result.length());
     return true;
```

```
}
else
    return false;
}
```

Summary

This chapter has provided you with methods to remotely program your Arduino board. You have seen in brief the process that takes place within the Arduino IDE for compiling, linking and uploading your sketch code on your favorite board. Based on the latter process you can either use the same tools with a ZigBee connection or use the Cloud-based web application for remotely programming your Arduino.

Next chapter closes this book with presenting various project ideas with things your can connect your Arduino to and manage data on the Cloud.

11 WHAT YOU CAN CONNECT TO THE CLOUD: PROJECT IDEAS

The previous chapters have provided you with great information on the Arduino and how you can connect it to the Cloud and share information coming from digital and analog sensors.

This chapter will discuss project ideas on what you could build and use it to communicate with the Internet of Things. Among others you will find out how you can send barcode and keypad data to the Cloud, manage online your home's power consumption and how to present data from the Cloud on your Arduino using LCD screens. You will also be briefly guided on how to create your own Arduino Shield for making any project circuit more compact and safer for deployment.

Within this concept, the chapter does not describe in details the implementation of the project ideas. Instead it describes the idea, provides some information regarding the circuit setup and code samples that will guide you through your own project implementation.

Connecting Things to your Arduino

Let's start by discussing on what you can connect to your Arduino and interact with its environment. By interaction we mean both getting inputs from the environment (e.g., sensing temperature and other or user actions like buttons, etc.) and generating outputs (e.g., controlling switches, etc.). Thus the suggested ideas are categorized based on sensors and actuators that can be connected to the Arduino. The following lists contain various sensors and actuators (some of them already presented in previous chapters) that have already been used by Arduino-fans in various projects. They can give you an idea of what else can be interfaced with Arduino and hopefully trigger your imagination for making your own projects.

Sensing Things and Inputs
While exploring the usage of Arduino in all kinds of projects you will find it also impossible to cover all the kinds of sensors that can be connected to the Arduino in a single book. So here is a short list with "Things" that sense and can be connected to your Arduino so that you can manage on the Cloud whatever information they produce.

Reading Barcodes
You can attach a barcode scanner directly to your Arduino and post readings to a Cloud service. A useful idea can by that each time you ran out

of supplies in your home or your office you scan the barcode of a product to be renewed. The code can go directly to an online database and this way you can make yourself a shopping list. You can also use a service like http://www.upcdatabase.com/itemform.asp and retrieve information automatically about the specific product. Then you access the list from your mobile phone the next time you go shopping and make sure you do not forget something! Using an Arduino Fio with a WiFi Bee shield you can integrate everything on a Barcode Scanner.

Barcode scanners usually come with a PS/2 connector (the one keyboards used to have before the USB). This type of connector has 5 or 6 pins that usually only 3 are used and are connected to Ground, input voltage (Vcc), and Data pins of a microcontroller.

The circuit for connecting the scanner with an Arduino Ethernet Shield can look like this:

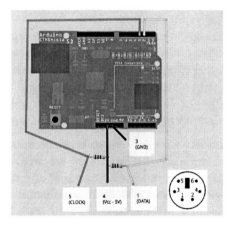

Figure 11-1. The Circuit Setup for connecting a Barcode Scanner With Your Arduino

You can use 2x22KOhm resistors as in displayed in the circuit. To read scanner values you can use the information and the code from here: http://arduino.cc/playground/ComponentLib/BarcodeScanner

Based on the code posted in the aforementioned URL, you can make the following changes to the loop() function in order to transmit the scanned code over the Internet to your Cloud application (e.g., hosted in Google App Engine) using a GET HTTP request and an Ethernet Shield:

Listing 11-1. Changes in the Loop Function of BarcodeScanner that enables it to send codes to a Web Application via GET requests.

```
void loop() {
  //Read the Barcode values and store them in dataValue
  variable
```

```
dataValue = dataRead();
// If there is a break code, skip the next byte
if (dataValue == SCAN_BREAK) {
    breakActive = 1;
}

// Translate the scan codes to numbers
// If there is a match, store it to the buffer
for (int i = 0; i < quantityCodes; i++) {
    byte temp = scanCodes[i];
    if(temp == dataValue){
        if(!breakActive == 1){
            buffer[bufferPos] = characters[i];
            bufferPos++;
        }
    }
}

}

// Send the buffer to the Cloud Application if SCAN_ENTER
   is pressed.
if(dataValue == SCAN_ENTER){
    if (client.connect()) {
        String s = String(bufferPos);
        client.println("GET /add?barcode=" + s + "
        HTTP/1.1");
        client.println("Host:barcodecloud.appspot.com");
        //here is your app engine url - app id with
        appspot.com
        client.println("Accept-Language:en-
        us,en;q=0.5");
        client.println("Accept-Encoding:gzip,deflate");
        client.println("Connection:close");
        client.println("Cache-Control:max-age=0");
        client.println();
        client.stop();
    }
    ...
}
```

and make sure you include the following in the beginning of your code:

```
#include <SPI.h>
#include <Ethernet.h>

byte mac[] = { 0xDE, 0xAD, 0xBE, 0xEF, 0xFE, 0xED };
byte ip[] = { 192, 168, 1, 200 }; // a valid IP on your
LAN
byte server[] = {72, 14, 204, 104}; // Google.com or
Appspot.com's IP
Client client(server, 80);
```

The Google App Engine application will have to contain a servlet that receive values through GET requests and saves them into the Google

Datastore. To do so it shall create a key and a Datastore Entity for setting the properties of the data you wish to store (e.g., the date the barcode arrives and the barcode itself). Then a DatastoreService is used to push new data into the entity. An example can be the following doGet() function as part of the main servlet:

Listing 11-2. The GET request implementation on the Servlet for storing Barcodes in the Google App Engine Datastore

```
public void doGet(HttpServletRequest req,
HttpServletResponse resp) throws IOException {

    String barcode = req.getParameter("barcode");
    Date date = new Date();
    Key BarcodeKey = KeyFactory.createKey("BarcodeDB",
    "BarcodeDB");
    Entity BarcodeData = new Entity("BarcodeData ",
    BarcodeKey );
    BarcodeData .setProperty("date", date);
    BarcodeData .setProperty("barcode", barcode );

    DatastoreService datastore =
    DatastoreServiceFactory.getDatastoreService();
    datastore.put(BarcodeData);

    //add any other functionality for the servlet
    . . .
}
```

For more details on the communication with a Cloud application hosted by Google App Engine you can check the related project in Chapter 7.

Logging Buttons Events on the Cloud

All kinds of buttons (check Chapter 5 for more information) can be interfaced with the Arduino. You could make your own doorbell system that apart from alerting you when a guest has arrived it could also log your visits online.

A more advanced system could be one that would have a keypad (see Figure 11-2) connected to the Arduino instead a single button. When user enters the correct code the Arduino can open a door or activate some system using a relay switch. You can manage the codes using a Cloud-based repository and also keep online logs of the access codes being used. The data encryption technique presented in Chapter 4 can be of great use here. Make sure you implement an administrator code that does not requite Internet connectivity (in case of network loss). You can find more details on how to interface and read codes from a keypad in this tutorial: http://www.arduino.cc/playground/Main/KeypadTutorial.

Figure 11-2. A Keypad that Can be Directly Interfaced With your Arduino (image courtesy of Sparkfun)

You can download and import the KeyPad library in your Arduino environment from the link above. Then you can modify the loop() function in one of the default examples as follows:

Listing 11-3. The modified loop() functions of the KeyPad Library Example that encrypts the entered code and sends it to a Web Application.

```
void loop(){
  char key = keypad.getKey();

  if (key != NO_KEY) {
    //add the entered key to th code string
    code[keycounter]=key;
    //increase the counter of the entered key strokes
    keycounter++;

  }
  //check if the complete code has been entered (in this
  //case code consists of 6 digits)
  if(keycounter==6) {
   //reset the counter
   keycounter = 0;
   //encrypt the code and send it over the Internet
   aes256_encrypt_ecb(&ctxt, (uint8_t*)code);
   if (client.connect()) {
    String s = String(code);
    client.println("GET /add?secretcode=" + s + "
    HTTP/1.1");
    client.println("Host:secretcode.appspot.com");
    client.println("Accept-Language:en-us,en;q=0.5");
    client.println("Accept-Encoding:gzip,deflate");
    client.println("Connection:close");
    client.println("Cache-Control:max-age=0");
    client.println();
    client.stop();
   }
  ...
}
```

The latter code will read keypad inputs until user has entered 6 of them (supposing that the correct combination consists of 6 digits). It encrypts then the code using AES256 encryption and sends it to a web application (secretcode.appspot.com) hosted by Google App Engine.

In addition you will need to add the following in the beginning of your sketch:

```
#include "aes256.h"
#include <SPI.h>
#include <Ethernet.h>

int keycounter=0;
char *code;
aes256_context ctxt;
byte mac[] = { 0xDE, 0xAD, 0xBE, 0xEF, 0xFE, 0xED };
byte ip[] = { 192, 168, 1, 200 }; // a valid IP on your LAN
byte server[] = {72, 14, 204, 104}; // Google.com or
                                      Appspot.com's IP
Client client(server, 80);
```

And the following inside the setup() function:

```
uint8_t key[] = {
    0x00, 0x01, 0x02, 0x03, 0x04, 0x05, 0x06, 0x07,
    0x08, 0x09, 0x0a, 0x0b, 0x0c, 0x0d, 0x0e, 0x0f,
    0x10, 0x11, 0x12, 0x13, 0x14, 0x15, 0x16, 0x17,
    0x18, 0x19, 0x1a, 0x1b, 0x1c, 0x1d, 0x1e, 0x1f
};

aes256_init(&ctxt, key);
```

The latter code (as you might remember from Chapter 04) contains the symmetric encryption key and also initializes the AES256 encryption algorithm. Your Cloud-based application will have to use the same key in order to successfully decode the messages, which in this case will be the key codes entered by users.

To do so in your Java Servlet code you will need initially to import the Java Cryptography Extension libraries that provide AES256 support.

```
import javax.crypto.Cipher;
import javax.crypto.KeyGenerator;
import javax.crypto.SecretKey;
import javax.crypto.spec.SecretKeySpec;
```

Then in the doGet() method of your servlet you need to implement the message decryption functionality using the same key Arduino uses for encryption.

Listing 11-4. The GET implementation of the Servlet Application that receives and decrypts the Keypad Code.

```
public void doGet(HttpServletRequest req,
HttpServletResponse resp) throws IOException {
    //Read the secret code Arduino sends through GET
    String code = req.getParameter("secretcode");

    //Start the decryption process
    try {
        byte[] sharedkey = {0x00, 0x01, 0x02, 0x03, 0x04,
        0x05, 0x06, 0x07,
        0x08, 0x09, 0x0a, 0x0b, 0x0c, 0x0d, 0x0e, 0x0f,
        0x10, 0x11, 0x12, 0x13, 0x14, 0x15, 0x16, 0x17,
        0x18, 0x19, 0x1a, 0x1b, 0x1c, 0x1d, 0x1e, 0x1f};

        SecretKeySpec skeySpec = new
        SecretKeySpec(sharedkey,"AES");

        Cipher cipher = Cipher.getInstance("AES");
        cipher.init(Cipher.DECRYPT_MODE, skeySpec);

        byte[] original =cipher.doFinal(code.getBytes());
        String originalString = new String(original);
    } catch (Exception e) {
        e.printStackTrace();
    }

    //Rest of Servlet implementation
    ...
}
```

The latter code receives the code as transmitted form the Arduino through the GET request into the variable 'secretcode'. It uses the same key and creates the cipher instance declaring 'AES' as encryption/decryption algorithm. The instance is initialized using the shared key and set to decryption mode. Then the encrypted code (in byte format) is decrypted and converted to a String variable again. You can use the variable to check that user is properly authenticated and return a flag (true or false) back to the Arduino.

Motion Sensors

There are several low-cost infrared or ultrasound sensors that can detect motion and interface directly with the digital ports of the Arduino. One potential usage of such sensors is to build a mini home-security system. When a sensor is activated (i.e. when some kind of motion is detected) you can have your Internet-enabled Arduino to send notifications to you. You can develop your own Cloud-based notification service as described in

Chapter 7 or use an existing service (like Nimbits as described in Chapter 9) and receive email, tweets or any other notification you might think of.

Pressure Sensors

You can use a pressure sensor which in fact is a force sensitive resistor (like the one in Figure 11-3) and be able to detect when you are running out of supplies. For example you can place the sensor under the coffee bag and be able to tell for its weight variation when it is time to replace it. By logging online the weight variations you can also tell how often coffee is made in your home or office.

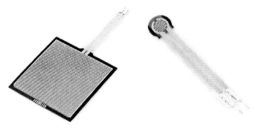

Figure 11-3. Various Force Resistive Resistors That Can Serve As Pressure Sensors (image courtesy of Sparkfun)

The force sensitive resistor can interface with your Arduino as a potentiometer. You can read its values through an analog pin like in Figure 11-4. Usually, the voltage read in the analog input of the Arduino is proportional to the inverse of the FSR resistance. This means that the more the weight on the sensor increases, the lower the values that you read.

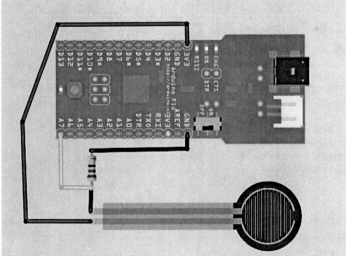

Figure 11-4. Connecting a Force Resistive Sensor with an Arduino Fio. The value of the resistor is 10K Ohm.

You might also want to implement a check mechanism that does not report false alarms when no weight is detected (object temporarily moved away) like the following:

Listing 11-5. The loop() function of the Arduino Sketch that reads input from the FSR Sensor and sends it to a Servlet on Google App Engine.

```
void loop() {
  //Read the value from the sensor:
  sensorValue = analogRead(sensorPin);
  boolean report = true;
  boolean isempty = false;

  //Check if the Sensor value is higher than a threshold
  //(e.g., 1000) that would indicate there is no weight on
  //it
  if(sensorValue > 1000) {
    //Add some delay so we avoid false alarms in case the

    //object is temporarily lifted
    delay(600000);
    //do not report this value
    report = false;
    //Check again to make sure the object is not empty or
    gone!
    sensorValue = analogRead(sensorPin);
    if(sensorValue > 1000) {
      isempty = true;
    }

  }
  else {
   report = true;
  }

  if(report) {
   //send the sensorValue to the Cloud
  }

  if(isempty) {
   //send an alert to the service
  }
...
}
```

The code will check if there is a high voltage reading in the analog input that would indicate very low pressure (low weight or no weight at all) at the FSR sensor. You can change the check threshold (1000 in this sketch) to meet your needs. In such a case, the sketch is delayed for 10 minutes so that to make sure the weight was not temporarily removed. Then it checks again and in case there is no weight again, then an empty alert is created.

Environmental Monitoring

We have already discussed a project that logs indoor temperature, light and humidity conditions through the Cosm Cloud service. There are several additional/alternative sensors you can use in order to manage online environmental and weather data. You can visualize the data using any of the Cloud-based services mentioned in the Book, like Cosm, Nimbits, ThingSpeak, etc.

Location Data

Using a GPS device or acquiring location through your Android phone and logging it online makes great sense when you are developing a 'wearable' project that monitors you on the move and/or the environment around you, since at the same time you can track the location of the sensor readings. For example you can create a project that uses a CO air sensor and can detect air quality levels (like one of the projects in Chapter 8) while you move or travel around. Then using a wireless network you can log the readings to your Cloud application and at the same time visualize your location using the coordinates and Google Maps.

In case you are creating a 'wearable' project (meaning that you carry the Arduino board and rest of components with you) instead of having it installed e.g., in a car, you might consider the communication with an Android phone a more convenient solution. The phone can provide you both with location data and Internet connectivity (WiFi or GPRS/3G/4G-based) and you can use your Arduino ADK board only for sensor reading.

However if you would still like to add GPS functionality to your Arduino (instead of using your Android phone) you can use a GPS Shield (like the one in Figure 11-5). The following code demonstrates how to receive location readings from the GPS module.

You need first to import the NewSoftSerial (http://arduiniana.org/libraries/newsoftserial/) and TinyGPS (http://arduiniana.org/libraries/tinygps) libraries into your Arduino environment.

Listing 11-6. Send GPS Information to A Cloud-based Application Hosted On Google App Engine

```
#include <NewSoftSerial.h>
#include <TinyGPS.h>
#include <SPI.h>
#include <Ethernet.h>

//Define the Ethernet related variables
byte mac[] = { 0xDE, 0xAD, 0xBE, 0xEF, 0xFE, 0xED };
byte ip[] = { 192, 168, 1, 2 }; // a valid IP on your LAN
```

```
byte server[] = {72, 14, 204, 104}; // Google.com or
Appspot.com's IP
Client client(server, 80);

// Define which pins you will use on the Arduino to
   communicate with your
// GPS module. In this sketch, the GPS module's TX pin will
   connect to the Arduino's RXPIN defined as pin 3.
#define RXPIN 3
#define TXPIN 2

// Create an instance of the TinyGPS object
TinyGPS gps;
// Initialize the NewSoftSerial library
NewSoftSerial uart_gps(RXPIN, TXPIN);

// This is where you declare prototypes for the functions
   that will be using the TinyGPS library.
void getgps(TinyGPS &gps);

// Define the variables that will be used
long latitude, longitude;
int year;
byte month, day, hour, minute, second, hundredths;

void setup()
{
  Ethernet.begin(mac, ip);
  //Initialize the GPS at the baud rate of the vendor
  uart_gps.begin(4800);

}

// The main loop of the code checks for data on
// the RX pin of the Ardiuno, makes sure the data is valid,
// it reads the GPS information
// and sends them to the Cloud service.
void loop(){
  // While there is data available
  while(uart_gps.available())
  {
      // load the data into a variable
      int in = uart_gps.read();
      // if there is a new valid sentence...
      if(gps.encode(in))
      {
        //Read the GPS Information into variables
        gps.get_position(&latitude, &longitude);

        gps.crack_datetime(&year,&month,&day,&hour,&minute,&
        second,&hundredths);
```

```
//Now Send the Data:
String lat = String(latitude);
String longi = String(longitude);
String date = String(year,DEC) + String(month,
DEC) + String(day, DEC) + String(hour, DEC) +
String(minute, DEC) + String(second, DEC) +
String(hundredths, DEC);
if (client.connect()) {
    client.println("GET/add?latitude="+lat+"&longitu
    de="+longi+"&date="+date+" HTTP/1.1");
    client.println("Host:gpsdata.appspot.com");
    client.println("Accept-Language:en-
    us,en;q=0.5");
    client.println("Accept-Encoding:gzip,deflate");
    client.println("Connection:close");
    client.println("Cache-Control:max-age=0");
    client.println();
    client.stop();
    }
  }
 }
}
```

Figure 11-5. A GPS Shield attached to an Arduino Board (image courtesy of Sparkfun)

The aforementioned code initially includes all the essential libraries for communicating with the GPS module and the Internet. Then it defines and initializes all essential variables for the Ethernet-based communication (like the MAC address, the IP of the Google App Engine, etc.) the TinyGPS object to be used the NewSoftSerial object that will communicate with the GPS module using the defined RX and TX pins.

The setup() function needs only to initialize the Ethernet module by binding the MAC with the IP address and initialize the GPS module. Most of the job is done inside the loop() function. While there are data available from the GPS module, it reads them and parses them into useful information (like latitude, longitude and date) using the TinyGPS library functions.

On-Body Sensors
Another great idea of what to send to the Cloud is to utilize it for monitoring vital signals using on-body sensors.

On-Body sensors use body features like temperature, skin resistance, light absorption/reflection by skin and others in order to detect and measure body temperature, heart pulses and more. There are quite a few commercial and/or open products that you can use with your Arduino in order to detect some of the latter bio signals and transmit them over the Cloud so that you can keep a track of your health state.

The Polar Heart Rate Module
You might be aware of the Polar Heart Rate Monitors that let you read your body's heart pulses using chest straps and wrist-watches (http://www.polarusa.com). Polar has also released a module that can be interfaced with any microcontroller and can give you access to the heartbeat readings of a chest strap. In order to interface it with your Arduino board all you need is the module itself and an external crystal oscillator at 32.76KHz. The module can be connected to a digital port of your Arduino and will give a HIGH state whenever there is a heart pulse detected.

The module is quite small and can be powered directly by your Arduino board. You can connect it with an Arduino Fio, add a WiFi Bee module and make yourself a mobile Wireless heart rate monitoring kit. You can check the circuit setup and the appropriate pin connections in Figure 11-6.

Check the code in the repository that demonstrates how to read heart beats, calculate an average heart rate and transmit it using the Ethernet Shield over the network to a Google App Engine-based servlet.

The idea behind the heart rate calculation is that the main execution loop waits until one pulse is detected. Then the current time is logged and a new internal loop is initiated until the second pulse is detected. Then the new time is logged and from the time interval an average rate for a 60 second period is calculated.

The latter code listing corresponds to an Ethernet based connection. Based on the projects of previous chapters you can easily adopt the sketch to support wireless communication using a WiFi Bee Shield.

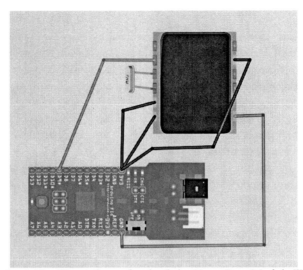

Figure 11-6. The Circuit Setup for the Polar Heart Rate Module With An Arduino Fio. The Module needs only Power and Ground connections and a 32.76KHz Crystal Oscillator.

The PulseSensor

A quite easier and affordable setup for sending your own heartbeats to the Cloud is the use of the PulseSensor (http://pulsesensor.com). PulseSensor is an open-source sensor that can be directly interfaced with the Arduino and allow users to measure their own heart pulses. It is also small and lightweight making it quite wearable and comes with an ear-clip accessory (see Figure 11-7).

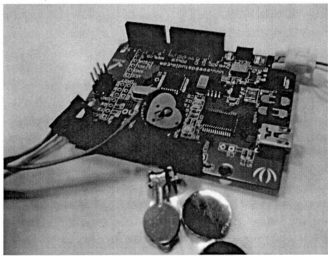

Figure 11-7. The PulseSensor. Small and lightweight can be worn as an ear-clip and let you read your heart pulses.

You can find the modified Arduino Sketch as provided by the developers of PulseSensor with additional Ethernet communication with a Google App Engine Application in the book's code repository.

The PulseSensor operates as an analog sensor and therefore is connected to an analog Arduino pin. It can operate in both 3.3V and 5V so it can directly powered by your board. The aforementioned code tries to identify the peaks in the analog input (called also as waveform) that correspond to heart beats.

For more information on how to setup and use the PulseSensor you can visit pulsesensor.com and look for the online manual.

Actuators

As presented in the various projects of this book, you can use actuators like relay switches or servo motors to control things like switches, turn on/off electrical appliances and even move objects like window shields.

An interesting application is to be able to control an actuator like a servo motor from a web-based application, like a java servlet on Google App Engine. In the following example the default Arduino Servo sample sketch is modified to receive commands from a Web application hosted on Google App Engine. The circuit setup is quite simple. As in Figure 11-8 you only need to power the Servo through your board and connect its input wire in digital pin 9.

Listing 11-7. The Arduino Sketch that moves a Servo based on values retrieved from a Servlet hosted in Google App Engine.

```
#include <Servo.h>
#include <SPI.h>
#include <Ethernet.h>

// Enter a MAC address and IP address for your controller
// below. The IP address will be dependent on your local
// network:
byte mac[] = {  0xDE, 0xAD, 0xBE, 0xEF, 0xFE, 0xED };
byte ip[] = { 192,168,1,2 };
byte server[] =  {72, 14, 204, 104}; // Google.com or
Appspot.com's IP
Client client(server, 80);

Servo myservo;  // create servo object to control a servo
String servoVal; //The value for moving the servo retrieved
from Cloud

char *val;    // variable to read the value from the
received Cloud string
```

```
int counter=0;   //needed for storing the Cloud string into
the val variable

void setup()
{
  myservo.attach(9);   // attaches the servo on pin 9 to the
                          servo object
}

void loop()
{

  if (client.connect()) {
    //You might want to pass a keycode parameter for
    //authenticating your Arduino first
    client.println("GET /add?keycode=123456789 HTTP/1.1");
    client.println("Host:mycloudservo.appspot.com");
    client.println();
    while (client.available()) {
      val[counter]=client.read();
      counter++;
    }
    client.stop();
    counter = 0;
  }

  myservo.write(atoi(val)); // sets the servo position
                              according to the Cloud value
  delay(15); // waits for the servo to get there
}
```

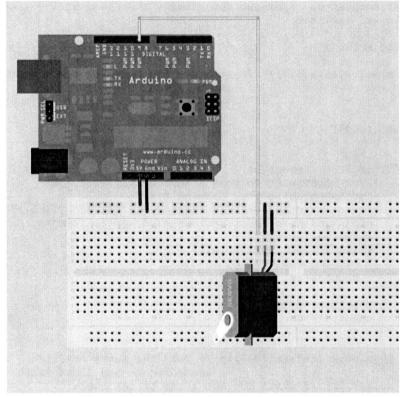

Figure 11-8. The Circuit Setup for connecting a Servo Motor with An Arduino Board

As with every provided example that relies on a Google App Engine Application you need to develop the appropriate servlet that in this case will send a variable in the range between 0-180 so that the servo is rotated.

The following sample code demonstrates the implementation of the doGet() function of such a servlet.

```
public void doGet(HttpServletRequest req,
HttpServletResponse resp) throws IOException {
    String mykey = req.getParameter("keycode");
    String ServoValue = "0";

    //Retrieve the Servo value from the DataStore:
    DatastoreService datastore =
    DatastoreServiceFactory.getDatastoreService();
    Key ServoDataKey = KeyFactory.createKey("ServoData",
    "ServoData");

    Query query = new Query("Servo",
```

```
ServoDataKey).addSort("id",
Query.SortDirection.DESCENDING);
List<Entity> readings =
datastore.prepare(query).asList(FetchOptions.Builder.wit
hLimit(1));
for (Entity reading : readings) {
    ServoValue = (String) reading.getProperty("value");
}

if(mykey.equals("123456789")) {
    resp.setContentType("text/plain");
    resp.getWriter().println(ServoValue);
}
...
}
```

The code receives a key so that only your Arduino can be authenticated. You can enhance the implementation to include encryption and decryption of the key for better security.

The servlet then assumes that the Servo value (that will make the servo move on the Arduino) is stored into the DataStore (through a different servlet for example). It retrieves it and it sends it back to the client as plain text.

Visualize Cloud Data on the Arduino

So far the majority of the projects described involve mostly the transmission of sensor data towards the Cloud and less the information retrieval from the Cloud. However, you might find useful sometimes that you retrieve such information on your Arduino. What can you do once you receive Cloud data on your Arduino? One idea is that you can visualize them on an LCD display.

There are several types of displays you can use, with variations on size, color and features, depending on what your project's needs are. To give you an idea the following demonstration is based on a 16x2 character monochrome LCD connected with your Arduino board according to the following setup:

* LCD RS pin to digital pin 12

* LCD Enable pin to digital pin 11

* LCD D4 pin to digital pin 5

* LCD D5 pin to digital pin 4

* LCD D6 pin to digital pin 3

* LCD D7 pin to digital pin 2

* LCD R/W pin to ground

A visualization of this setup is presented in Figure 11- 9. You can also use a potentiometer to adjust the brightness level of the display.

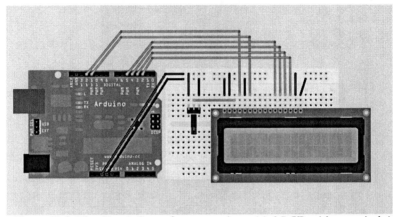

Figure 11-9. The Circuit Setup for connecting a 16x2 LCD with your Arduino Board. The circuit contains also a potentiometer for setting the LCD brightness level.

The Arduino Environment comes with an appropriate LCD Library (compatible with the majority of LCD vendors, make sure yours is compatible before using the library) preinstalled (named as 'LiquidCrystal') so that you can use it directly in your sketches.

Consider that you would like to display on the LCD information from sensor feeds as stored in a Cloud service like ThingSpeak. Even if you do not use ThingSpeak for managing you sensor data online, ThingSpeak offers a list with public channels that users can browse and retrieve the latest sensor readings. You can find the list here: https://www.thingspeak.com/channels/public.

As an example you can use any public channel you wish and try to retrieve the CSV representation of the sensor information.

Listing 11-8. The Arduino sketch that displays sensor information from ThingSpeak Channels on LCD.

```
#include <LiquidCrystal.h>
#include <SPI.h>
#include <Ethernet.h>

// Enter a MAC address and IP address for your controller
below.
// The IP address will be dependent on your local network:
byte mac[] = {  0xDE, 0xAD, 0xBE, 0xEF, 0xFE, 0xED };
byte ip[] = { 192,168,1,2 };
```

```
byte server[] = { 184,106,153,149}; // ThingSpeak

// Initialize the Ethernet client library
// with the IP address and port of the server
// that you want to connect to (port 80 is default for
//HTTP):
Client client(server, 80);

// initialize the library with the numbers of the interface
//   pins
LiquidCrystal lcd(12, 11, 5, 4, 3, 2);

void setup() {
  // set up the LCD's number of columns and rows:
  lcd.begin(16,2);
  //Initialize the Ethernet module
  Ethernet.begin(mac, ip);
  delay(1000);
}

void loop() {
  // set the cursor to (0,0):
  lcd.setCursor(0, 0);

  // set the display to automatically scroll:
  lcd.autoscroll();
  if (client.connect()) {
    client.println("GET /channels/277/field/1.csv
    HTTP/1.0");
    //Tell the Web Server to look for the
    //api.thingspeak.com URL
    client.println("Host:api.thingspeak.com");
    client.println();

    //Read input from client (the csv output) and print it
    //on LCD
     while (client.available()) {
      lcd.print(client.read());
     }
   client.stop();
  }

  delay(60000);

  // clear screen for the next loop:
  lcd.clear();
}
```

The sketch above connects the Arduino (through the Ethernet Shield) to the ThingSpeak API and makes a GET request for channel 277 and its CSV data representation (1.csv). The output is clean text, comma separated (as it should since you asked for CSV format) that contains the last sensor readings. While characters are arriving from the web page they are also displayed on the LCD screen.

Monitor Power consumption on the Cloud

On of the most popular uses of Arduino is for monitoring power consumption of electrical appliances or in total for a small unit like a house or a small office. Cosm currently serves more than 250 feeds that are related to energy consumption (you can search for feeds tagged with 'kWh').

Power consumption involves the measurement of the current that is consumed by a device and the measurement of the voltage applied on the device. The result is usually the product of the two.

The most common way to measure the current consumed by a device is to apply a CT sensor on the wiring that carries the current to the device. A CT sensor (see Figure 11-10) is a current transformer (this is where the abbreviation CT comes from) that is placed around the wire and produces a reduced current proportional to the current that flows into the wire. By knowing the characteristics of the CT sensor and by doing some math one can calculate the current consumption with great accuracy. Since the output of the sensor is reduced current, it can also be measured by your Arduino making yourself a nice power meter for your electrical appliances or even your entire home.

If you are interested into building such a system there is no need to worry about how to build the circuit and implement the math behind it. People behind openenergymonitor.org have created an open-source project that provides you with all the information you need and an Arduino library for building your own power meter.

Let's consider the following setup that will enable you to measure your home's power consumption and monitor it online using a Cosm feed.

Parts needed
To build yourself your own power meter and log energy consumption on a Cloud application you will need the following:
- 1x Arduino Ethernet or an Ethernet shield for measuring power and sending it over Ethernet network to Cosm
- 1x 9V AC-AC Power Adapter in order to measure the voltage
- 1x 100kOhm resistor
- 5x 10kOhm resistors
- 1x 10uF capacitor
- 1x CT sensor
- 1x 46Ohm resistor

For more information on the latter parts and why they are needed you can look at the OpenEnergyMonitor website.

Figure 11-10. A Current CT Sensor (image courtesy of Seeedstudio)

Prepare Cosm for Managing Power Feeds

In order Cosm to receive and visualize your power consumption you need to set up the appropriate feed and datastream following the instructions given in Chapter 8 for the use of Cosm service. Make sure you note the master key of your feed and your feed's ID.

Regarding datastreams, you can create three of them. One for monitoring the real power consumption, one for the apparent power and one for the power factor as calculated by the Emon library respectively.

Circuit Setup

Connect the capacitors and the resistors with your Arduino board, the power adapter and the CT sensor as indicated in Figure 11-11.

The Code

For this project you will need to import the Emon Library into your Arduino Environment. You can download the library from here: https://github.com/openenergymonitor/MainsACv3

The following code is adopted from the default library example that reads voltage and current readings from analog pins 4 and 3. It has been adopted to transmit the data over Ethernet to a Cosm feed.

Listing 11-9. Using the Emon Library to transmit Power Monitoring Data to Cosm Feed

```
//----------------------------------------------------------
-------------------
// Mains AC Non Invasive 3
// Last revision 30 November 2009
// Licence: GNU GPL
// By Trystan Lea
```

```
//------------------------------------------------------------
------------------

//------------------------------------------------------------
------------------
// Load Energy Monitor library and create new instance
//------------------------------------------------------------
------------------

/*
Modified by Charalampos Doukas
to send power monitoring data
to a Cosm Feed using the Ethernet Library
*/

#include <SPI.h>
#include <Ethernet.h>
#include "Emon.h"     //Load the Emon library

EnergyMonitor emon;   //Create an instance

//Set up ethernet and network related data
byte mac[] = { 0xCC, 0xAC, 0xBE, 0xEF, 0xFE, 0x91 };
byte ip[] = { 192, 168, 1, 2 }; // no DHCP so we set our own
IP address
byte CosmServer[] = { 173, 203, 98, 29 };

//create a Client object for handling connection
Client localClient(CosmServer, 80);

//Useful char buffers for creating the CSV datastream
content
char buf1[16];
char buf2[16];
char buf3[16];
char DataBuff[16];

// power variables
float realPower;
float apparentPower;
float powerFactor;

//feed data function, provides data for the POST command in
comma separated values (CSV)
void feedData()
{
   ftoa(buf1, realPower, 2);
   ftoa(buf2, apparentPower, 2);
   ftoa(buf3, powerFactor, 2);
   //we save all variables into one char variable DataBuff
   sprintf(DataBuff,"%s,%s,%s", buf1, buf2,buf3);
}
```

```
//Define the sendData() function that handles the
communication with Cosm implementing the protocol for
sending feeds based on Cosm API v2.
void sendData(){
  if (localClient.connect()) {
    feedData();
    int content_length = strlen(DataBuff);
    localClient.print("PUT /v2/feeds/");
    localClient.print("12345"); //Put your feed ID
    localClient.print(".csv HTTP/1.1\nHost:
    api.pachube.com\nX-PachubeApiKey: ");
    localClient.print("NkyX--------------------------------
    90s"); //Replace with your Cosm API Key
    localClient.print("\nUser-Agent: ");
    localClient.print("\nContent-Type: text/csv\nContent-
    Length: ");
    localClient.print(content_length);
    localClient.print("\nConnection: close\n\n");
    localClient.print(DataBuff);
    localClient.print("\n");
  }
}

//------------------------------------------------------------
--------------------
// Setup
//------------------------------------------------------------
--------------------
void setup()
{
  emon.setPins(4,3); //Energy monitor analog pins
  //Energy monitor calibration
  emon.calibration( 1.116111611, 0.128401361, 2.3);
  //initiate Ethernet module
  Ethernet.begin(mac, ip);
}

//------------------------------------------------------------
--------------------
// Main loop
//------------------------------------------------------------
--------------------
void loop()
{
  emon.calc(20,2000); //Energy Monitor calc function
  realPower = emon.realPower;
  apparentPower = emon.apparentPower;
  powerFactor = emon.powerFactor;
  //send the data over Cosm
  sendData();
```

```
  delay(10000);
}
//----------------------------------------------------------
------------------
```

```
//Convert double to char (due to currently sprintf in
Arduino fails to do so)
char *ftoa(char *a, double f, int precision)
{
  long p[] =
  {0,10,100,1000,10000,100000,1000000,10000000,100000000};
  char *ret = a;
  long heiltal = (long)f;
  itoa(heiltal, a, 10);
  while (*a != '\0') a++;
  *a++ = '.';
  long desimal = abs((long)((f - heiltal) * p[precision]));
  itoa(desimal, a, 10);
  return ret;
}
```

By examining the sketch code above you will figure out that it uses the philosophy of the projects presented in Chapter 8.

Other Ways to Visualize Online Power Data

People behind the OpenEnergyMonitor have also developed and released a very powerful and open-source web-app for processing, logging and visualizing energy, temperature and other environmental data. They call it "EmonCMS" (see Figure 11-12) and you can get it from here: http://openenergymonitor.org/emon/emoncms

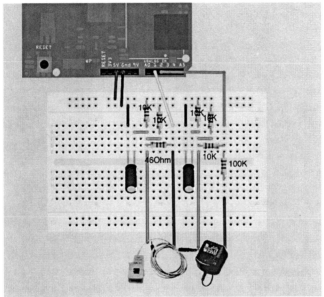

Figure 11-11. The Circuit Setup for the Energy Monitor Project.

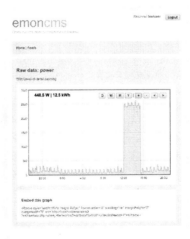

Figure 11-12. The emonCMS Open Source Energy Visualization Content Management System

Create your own Arduino Shield

Given that you have followed the instructions of the book for the projects in each chapter, you must have realized the great usability of the breadboard. For sure it is a quick and convenient way (avoiding procedures like cutting wires and soldering pins) for connecting various parts and your Arduino together. However the breadboard is only good for prototyping. When you have completely tested your circuit setup and your Arduino code, leaving your project running on the breadboard is not safe (jumper wires can be accidentally removed quite easily) and also requires from you to buy yourself a new breadboard each time you want to build/test something new.

So what is the best solution for your project that would also require the minimum effort for connecting and soldering components together? To make your own Arduino Shield! Making an Arduino Shield includes the process of designing a PCB board (the board that holds together the ATMega microcontroller and the rest of the components that make your Arduino) using specific layout and design instructions (like placing Arduino pin headers, etc.) and of course to develop one. You can either develop the board yourself (which requires a lot of effort and skills) or order one online from various PCB service providers (like http://batchpcb.com/) by providing the appropriate design files. Fortunately the process of designing a PCB is much more easy than it sounds using the appropriate tools. On such tool is the 'Fritzing' (http://fritzing.org) open source electronic design automation tool.

Fritzing includes graphical design environment (see Figure 11-13) where users can simply design electronic circuits the way they build their prototypes using an Arduino, a breadboard and various electronic components. It offers mainly three different views you can use for designing the circuit. The first and default one is the breadboard view, the second one is the schematic view (where you can see how the various components are connected to the I/O pins of the microcontroller) and the third one is the PCB view.

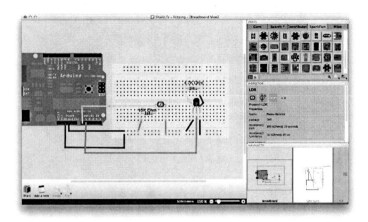

Figure 11-13. The main interface of Fritzing for designing circuits in Breadboard view

To design a circuit setup users simply select the components from the 'Parts' list on the right of the graphical interface and drag them on the breadboard area. They make the connections by dragging virtual links between the pins of the components. Fritzing provides useful guidance for connections by highlighting the connected pins on the breadboard.

When the circuit design is completed, you can switch to the PCB view and see how the components can fit on a PCB board and customize it by selecting different layer types, resize it and/or rearrange the components. When you need to build yourself an Arduino shield, Fritzing offers three great features:

1. It allows you to use an Arduino Shield layer (see Figure 11-14) which includes all the additional essential components (e.g., the Arduino I/O pin extension headers) placed appropriately on the shield.

2. It has an advanced connection design feature called 'Autoroute'. The latter will make any necessary connections or re-arrange existing ones so that there is no overlapping between them.

3. It has an export feature that generates all the appropriate files for developing the PCB board with one single click. You do not need to design anything else yourself; you just use these files to order your PCB using an online service.

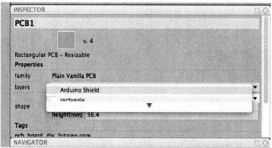

Figure 11-14. Selecting the Arduino Shield as a PCB Layout in Fritzing.

Figure 11-15 presents the PCB layout as generated automatically by Fritzing for Project 1 of Chapter 8 (measuring temperature, light and humidity using a LDR and a DHT22 sensor). The layout corresponds to an Arduino Shield that can be placed over an Arduino Ethernet module or an Ethernet Shield and make a compact solution for sending the sensed data to Cosm Service.

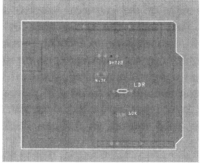

Figure 11-15. A PCB Design using Fritzing for an Arduino Shield that implements the Circuit for measuring indoor Light, Temperature and Humidity using a LDR and a DHT22 Sensor.

Summary

You have been presented with some ideas of what you can connect to your Arduino in terms of sensor inputs and outputs and how the latter can interact with Cloud applications. You have also seen another great feature of the Arduino in combination to the power of open source software; how to design and make your own Arduino Shields for implementing your project circuits on PCBs and making them last longer.

This final Chapter of the book aims to trigger your imagination about what you can make with your Arduino and an Internet connection. Your imagination can be the only limit!

Make sure you visit frequently the book's blog at **http://www.buildinginternetofthings.com** for updates about the code, projects, Arduino and IoT platforms!

ABOUT THE AUTHOR

Charalampos Doukas is an Information Systems Engineer who spends most of his time doing research on medical sensors and medical data. His first encounter with Arduino was back in 2007 when he was looking for open wireless sensor platforms. Since then he has used the Arduino to build several monitoring and home automation projects. His interests have expanded in Cloud computing and Internet of Things platforms. He has given several lectures and tutorials on the Arduino and on communication with Cloud systems. Charalampos has authored and co-authored several research papers on scientific journals and book chapters.

CPSIA information can be obtained at www.ICGtesting.com
Printed in the USA
LVOW102007270613

340559LV00022B/1185/P